D0620848

METHANE GENERATION AND RECOVERY FROM LANDFILLS

by

EMCON ASSOCIATES
1420 Koll Circle
San Jose, California 95112

for

CONSOLIDATED CONCRETE LIMITED

and

ALBERTA ENVIRONMENT

Second Printing, 1982

P.O. Box 1425, 230 Collingwood, Ann Arbor, Michigan 48106

Library of Congress Catalog Card Number 80-67725
ISBN 0-250-40360-9

Manufactured in the United States of America

Butterworths, Ltd., Borough Green
Sevenoaks, Kent TN15 8PH, England

ACKNOWLEDGMENTS

The authors wish to express their appreciation to the following reviewers for their editorial and technical comments:

Dr. Foppe DeWalle, Department of Environmental Health, University of Washington, Seattle, Washington

Mr. William G'Sell, University of Washington, Seattle, Washington

Dr. James O. Leckie, Department of Civil Engineering, Stanford University, Stanford, California

Mr. Max Blanchet, Pacific Gas and Electric Company, San Francisco, California

Mr. Richard T. Mandeville, Gas Recovery Systems Inc., Pasadena, California

Mr. Charles Moell, Alberta, Canada

In addition, the authors wish to thank the following people for their valued comments and suggestions:

Dr. James Roberts, Department of Civil Engineering, California State University at San Jose

Dr. Donald L. Wise, Dynatech R/D Company, Cambridge, Massachusetts

CONTENTS

Illustrations

INTRODUCTION

Until quite recently, society has disposed of household waste by open dumping, frequently followed by burning to reduce volume and produce heat. Burning reduces the volume of refuse principally by accelerating the decomposition of the organic wastes in the presence of oxygen. Since the late 1960's and early 1970's, however, sanitary, or environmentally managed, landfilling of our household and municipal wastes has replaced the ancient practice of dumping and burning. Developed in response to public concern for the protection of the environment, sanitary landfilling involves covering the refuse daily with soil, thereby creating anaerobic conditions for refuse decomposition.

A significant by-product of the sanitary landfill has been the creation of conditions suitable for the decomposition of organic wastes to produce a mixture of gases whose major component is methane--the primary constituent of natural gas. In essence, the degradation process taking place in a landfill is analogous to that occurring in a sewage sludge digestor; the major difference is that optimum conditions for methane production are rarely, if ever, encountered in a landfill.

In recent years there has been a growing interest in the generation of landfill gas. Although the main reason for this interest is the scarcity of energy sources and the rising prices of natural gas and fossil fuels, there are several other contributing factors. First, the trend toward regionalization of disposal sites has resulted in the development of landfills in large metropolitan areas where substantial quantities of refuse can be deposited to great depths. Since methane production is a function of the quantity of waste deposited (as well as a number of other factors such as temperature, moisture content, and pH), large regional landfills represent prime candidates for gas recovery. In excess of 5 million cubic feet per day of equivalent pipeline standard gas (1000 Btu/scf) may be recoverable from each of the largest landfills.

Another development which has contributed to the interest in landfill gas generation is the emergence in the United States of regulatory standards to control the hazard posed by migrating methane gas. In concentrations between 5 and 15% by volume in air, methane is flammable at atmospheric pressure and ordinary temperatures. The potential for hazard is heightened by the ease with which methane may migrate subterraneously, often to significant distances through permeable media; public safety may be endangered if methane accumulates in a poorly ventilated area and subsequently achieves combustible concentrations. Solid waste disposal guidelines mandated by Section 1008 of the Resource Recovery and Conservation Act of 1976 (RCRA), slated for final publication in early 1979, establish criteria for control of landfill gas migration. The proposed criterion for explosive gases stipulates that concentrations of such gases must remain below 25% of the lower explosive limits (LEL) in facility structures, and below their LEL in the soil at the property boundary.

The forthcoming gas control criterion is expected to result in the need for gas migration control systems at numerous landfills throughout the country. Where active control systems are required, the gas-gathering network often can be adapted to gas recovery for little additional capital investment, thereby creating the basis for a profitable gas recovery project. The recovered methane might be used on-site for heat or power generation, sold as a fuel to nearby industrial customers, or, in some cases, sold to utilities for subsequent use by private customers.

The increasing interest on the part of regulatory agencies and private research groups in the generation and commercial recovery of landfill methane gas has created the impetus for this state-of-the-art study of the theories and methodology related to landfill gas recovery. Although the major focus of this state of the art of landfill gas is on theory and methodology, it is important to note that the successful or commercial application of this information requires an understanding of a number of additional disciplines, such as:

1. Landfill design and operation, including control systems for leachate and gas;
2. Economic feasibility studies of selected landfills;
3. Regulatory constraints on the commercial sale of recovered landfill gas;
4. Legal arrangements between the landfill owner and prospective energy user(s); and
5. The transmission of gas or energy.

The general approach used in this study was to (1) gather, collate and evaluate available literature pertinent to landfill gas production and recovery, and (2) summarize this information, emphasizing the currently used theories and approaches rather than attempting a comprehensive review of methodology in the field. Additional input to the study came from discussions with experts in the area of gas generation and recovery. The information presented is organized as follows: (1) a conceptual review of the methane fermentation processes; (2) discussion and summary of data on the composition of refuse materials; (3) procedures for estimating the theoretical maximum methane yield to be expected from composite refuse; (4) review and discussion of the available models for the time-dependency of gas production; (5) gas flow in landfills; (6) field testing of landfills to determine gas recovery potential; (7) methods of gas recovery; and (8) processing and utilization of the gas.

METHANE FERMENTATION PROCESSES

In the methane fermentation process, stabilization of organic wastes is carried out by microorganisms under anaerobic conditions, resulting in the production of methane and carbon dioxide. Understanding the basic kinetic parameters which affect the degree of waste degradation (stabilization) should aid in appreciation of the overall rate-controlling factors as they occur in a landfill context. Most of the literature cited in this section reflects work done on the anaerobic liquid waste treatment process, but the basic considerations are also valid for the landfill context.

GENERAL MICROBIOLOGY AND BIOCHEMISTRY

In general, any waste material susceptible to aerobic degradation can also be treated anaerobically. One of the few exceptions to this statement is lignin, an amorphous polymeric substance in wood that binds the cellulose fibers. Lignin comprises about 30% of wood, while cellulose-type fibers account for 55 to 75% of the wood material. Unlike lignin, cellulose materials are more readily degradable by anaerobic than by aerobic processes. Wood-derived paper product wastes comprise the bulk of the moderately decomposable wastes in a landfill. The biodegradability of wood products is discussed in some detail in a later section.

Anaerobic decomposition of complex organic materials is normally considered to be a two-stage process, as indicated in Figure 1 [28].* In the first stage, there is no methane production. In this stage, the complex organics are altered in form by a group of facultative and anaerobic bacteria commonly termed the "acid formers". Complex materials such as cellulose, fats, proteins, and carbohydrates are hydrolyzed, fermented, and biologically converted to simple organic materials [20, 21].

---

* Recent research on anaerobic processes has resulted in some modification of the theory presented on this section (p.5, para. 3 to p.9, para. 1). See Appendix D for a synopsis.

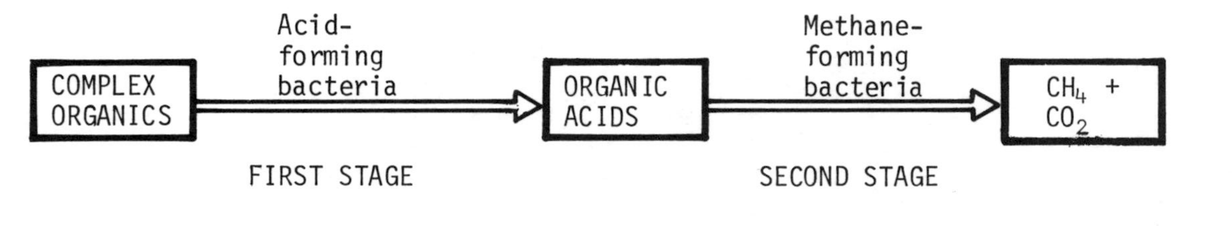

Figure 1. Two stages of anaerobic decomposition of complex organic wastes.

For the most part, the end products of this first-stage conversion are organic fatty acids. Acid-forming bacteria bring about these initial conversions, obtaining for growth the small amounts of energy released; in the process, a small portion of the organic waste is converted to bacterial cells. This first stage of biodegradation is required to place the organic matter in a form suitable for the second stage of decomposition.

During the second stage of methane fermentation, the organic acids are consumed by a special group of methanogenic bacteria and converted into methane and carbon dioxide. The methanogenic bacteria are strictly anaerobic, and even small quantities of oxygen are toxic to them.

Figure 2 illustrates the pathways for the fermentation of complex wastes such as sewage sludge, food wastes and vegetable matter (e.g., garden wastes, paper, wood) [28]. The numerical values shown represent the portion of waste chemical oxygen demand (COD) which is converted by the various routes. Not shown in the figure is the portion of the various organics that is converted to microbial cells as the waste is fermented. This portion may be as high as 20% during the fermentation of carbohydrates, and as low as 4% during methane fermentation of the saturated organic acids.

Figure 2 indicates that, during the complete methane fermentation of complex wastes, over 70% of the methane results from the fermentation of acetic acid. Also, about 30% of the waste COD must pass through propionic acid on its way to becoming methane. The importance of these two acids, and of the microorganisms which consume these acids in the formation of methane, is thus indicated.

Waste stabilization requires a balance among all the organisms; the establishment and maintenance of this balance is normally indicated by the concentration of volatile acids and pH in the system [28]. When the system is in balance, the methanogenic bacteria use the acid intermediates as rapidly as they are produced. However, if the methanogenic bacteria are not present in suitable numbers, or are being slowed down by unfavorable environmental conditions (e.g., pH, toxic materials, etc.), they will not use the organic acids as rapidly as those intermediate products are produced by the acid formers. As a result, the volatile acids will increase in concentration [28].

During start-up of the methane fermentation process, when excessive sudden shock loads are added, when temperature fluctuations occur, or when inhibitory materials enter the system, the balance between methane-producing and acid-forming organisms may be upset, leading to an organic acid increase and a drop in pH. If the pH decrease is not prevented (i.e., if the pH drops

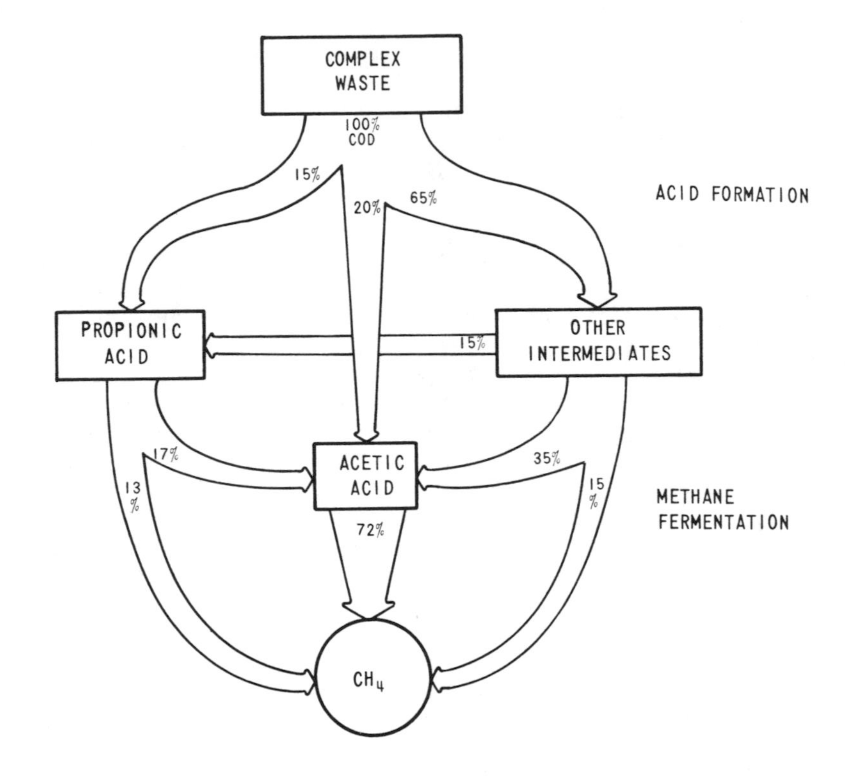

Figure 2. Pathways for methane fermentation of complex wastes.

below 6) during this period, the methane-fermenting organisms may be destroyed, thus causing a severe and costly setback in the methane production process. To prevent such a setback, and to aid in optimizing the process, it would be desirable to have the ability to add alkaline materials in order to maintain the process balance. A monitoring system for determination of acidity and pH in the landfill leachate would be an additional asset for management of the system's environment.

PROCESS BIOCHEMISTRY

A generalized reaction for the overall methane fermentation of a waste with an empirical chemical formulation of

$$C_nH_aO_bN_c$$

to methane, carbon dioxide, and bacterial cell ($C_5H_7O_2N$) is as follows:

$$C_nH_aO_bN_c + (2n + c-b - \frac{9sd}{20} - \frac{de}{4} H_2O =$$

$$(\frac{de}{8})\ CH_4 + (n-c - \frac{sd}{5} - \frac{de}{8})\ CO_2 + \frac{sd}{20}\ C_5H_7O_2N$$

$$+ (c - \frac{sd}{20})\ NH_4^+ + (c - \frac{sd}{20})\ HCO_3^- \qquad (1)$$

where,

$d = 4n + a - 2b - 3c$
s = the fraction of waste chemical oxygen demand (COD) synthesized or converted to cells
e = the fraction of waste COD converted to methane gas for energy

and

$$s + e = 1. \qquad (1a)$$

The value of s varies with the waste composition, the average solids retention time in the system (θ_c), and the cell decay rate (f), as follows:

$$s = a_e \left[\frac{(1+0.2f\ \theta_c)}{(1+f\ \theta_c)}\right] \qquad (2)$$

where,

θ_c = solids retention time, days

$a_e = s_{max}$ when $\theta_c = 0$
f = cell decay rate, day^{-1} (per day).

The value 0.2 represents the refractory portion of the bacterial cells formed during cell decay. Values for a_e, the maximum value for s obtained when θ_c equals zero, are listed in Table 1 for various components of wastes; carbohydrates give the maximum yields and fatty acids, the minimum [31].

TABLE 1. Values for a_e and Y for methane fermentation of various wastes components (Ref. 31)

Waste Component	Chemical Formula	a_e	Y(Organism Yield Coefficient) gm cells per gm COD consumed
Carbohydrate	$C_6H_{10}O_5$	0.28	0.200
Protein	$C_{16}H_{24}O_5N_4$	0.08	0.056
Fatty Acids	$C_{16}H_{32}O_2$	0.06	0.042
Domestic Sludge	$C_{10}H_{19}O_3N$	0.11	0.077
Ethanol	C_2H_6O	0.11	0.077
Methanol	CH_4O	0.15	0.110
Benzoic Acid	$C_7H_6O_2$	0.11	0.077

The specific maintenance rate, f, has also been termed more generally the cell decay rate, as decrease in biomass may occur through death, lysis, endogenous metabolism, and predation, as well as through energy utilization for maintenance [23]. Typical values for f have ranged from 0.01 per day to 0.05 per day in mixed cultures. Organism decay rates under anaerobic conditions tend to be lower in general than under aerobic conditions; f ranges from 0.02 to 0.06 for aerobic systems and 0.01 to 0.04 for anaerobic systems [23].

The biological yield factor, Y, generally ranges from 0.03 to 0.15. For anaerobic systems, Y is 0.03 to 0.05 for protein and fatty acid substrates and 0.14 to 0.16 for carbohydrate wastes.

pH AND ALKALINITY REQUIREMENTS

The optimum pH for methane fermentation is about 7 or slightly above, but the methane bacteria generally are not harmed unless the pH drops below the range 6.0 to 6.5. The lower the pH, the shorter the time for a given decrease in activity. For this reason, it is essential that the pH be maintained above 6 at all times in those cases where pH control is possible. When pH control is not possible, the process will probably not be operating under optimum conditions.

Figure 3 illustrates the relationship between pH, bicarbonate concentration and carbon dioxide partial pressure in an aqueous, anaerobic system [28]. When the bicarbonate alkalinity drops below about 500 mg/L, and with the normal percentage of carbon dioxide in the gas (e.g., about 50%), the pH will drop dangerously close to 6.0. Since most landfills will start out under acid conditions (pH between 5 and 6), they will not initially provide optimal pH conditions for maximum methane production unless the pH is controlled externally.

When anaerobic systems become unbalanced, the volatile acid concentration increases, destroying the bicarbonate alkalinity according to the following reaction:

$$HCO_3^- + HAc = H_2O + CO_2 + Ac^- \qquad (3)$$

where,

HCO_3^- is the bicarbonate ion and HAc is acetic acid.

This results in a pH decrease as indicated in Figure 3. The total alkalinity will not change greatly, as Ac^- also is measured as alkalinity. When the volatile acid concentration and the total alkalinity (usually titrated to pH 4.0) are known, then the bicarbonate alkalinity can be computed by the following relationship:

$$BA = TA - 0.71\ (VA)$$

where,

BA = bicarbonate alkalinity, mg/L as $CaCO_3$
TA = total alkalinity, mg/L as $CaCO_3$
VA = volatile acids, mg/L as acetic acid (HAc)

According to Dr. DeWalle of the University of Washington (pers. comm.), bicarbonate alkalinity in landfills tends to be relatively low, since the concentration of inorganic carbon in leachate is typically less than 50 mg/L. However, the release of ammonia from nitrogenous organics in refuse is thought to

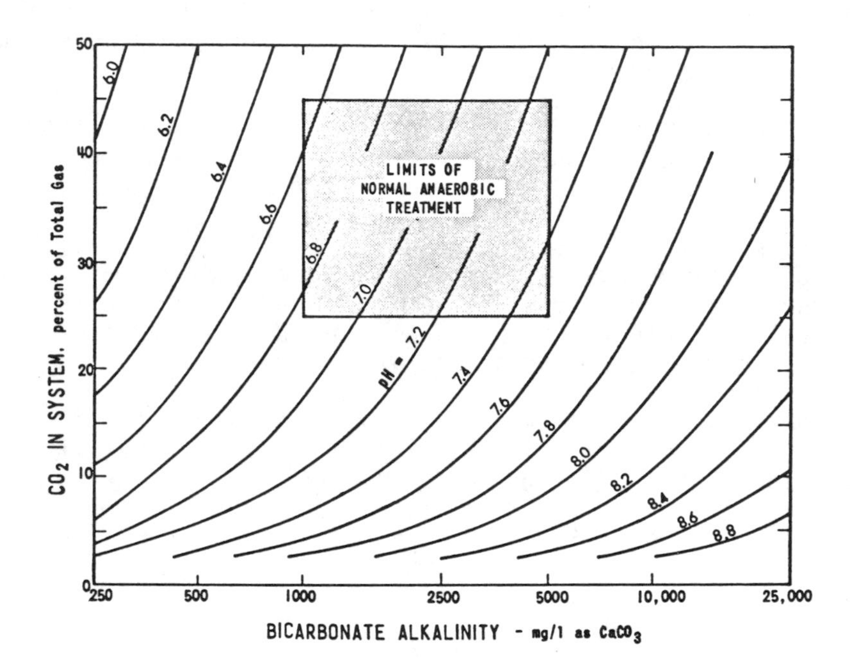

Figure 3. Relationship between bicarbonate alkalinity, pH, and carbon dioxide percentage in anaerobic treatment.

counteract the drop in pH according to the following reaction:

$$HAc + NH_3 = NH_4^+ = NH_4^+ + Ac^-$$

where,

HAc is acetic acid
Ac^- is the acetate ion
NH_4^+ is the ammonium ion.

Note that the above reaction neutralizes the organic acids to the extent that the ammonium ion is produced.

SUMMARY

Although the activities and very existence of a microbial population are associated with a multitude of abiotic factors, a range (maximum and minimum) can usually be established for each of these factors. The extremes of each range are physiological limits beyond which the microbial population is unable to maintain itself or to perform a vital function. For each factor, an optimal level or range usually can be established, in addition to an upper or lower limit.

In a heterogeneous mixed-media system such as is found in a landfill, it is to be expected that many microenvironments, exhibiting varied conditions, may be formed at different locations within the refuse and may support the growth of widely varying types of organisms. For example, one may find methane produced in a landfill which appears to have a leachate pH of 5 (even though methanogenic bacteria cannot function at pH 5). This unusual situation can occur because there are pockets or regions (microenvironments) in the landfill where conditions allow these organisms to survive.

COMPOSITION OF MUNICIPAL REFUSE

GENERAL CONSIDERATIONS

Refuse composition directly affects both the total yield and the rate of methane production during anaerobic decomposition in a landfill. Methane production is stimulated by a waste having a high percentage of biodegradable organic materials (food and garden wastes, paper, textiles, and wood). To facilitate estimating the overall rate of methane production as a function of time, the biodegradable organics in refuse can be subdivided into categories; for example, rapidly decomposable (food and garden wastes), and moderately decomposable (paper, textiles, and wood).

Estimating the bulk composition of municipal refuse is a task requiring attention to questions of (1) sample size, (2) frequency of sampling, and (3) total number of samples [51]. Since municipal refuse is a composite waste material made up of particles or pieces the sizes of which range over several orders of magnitude, the most reliable and accurate estimates of composition have been based on sample masses of at least 90.7 kg [7].

COMPOSITION AND CHEMICAL ANALYSIS OF REFUSE

Table 2 summarizes the best available literature on the composition of residential and commercial refuse. Data are given as the percent total of total wet weight of refuse. Data indicate that approximately 70 to 80% of the total wet weight is comprosed of organic-based materials.

A high refuse moisture content (in the range of 60 t0 80%--wet weight) favours maximum methane production. However, the moisture content of refuse at the time of placement is normally well below this range, averaging about 25%. Typical moisture

Table 2. Municipal refuse composition (percent on wet weight basis).

REFUSE COMPONENT	REFERENCE NUMBER[a]													
	7	16	17	39	40	43	47	47	47	47	48[b]	49	49	54
Food Waste	10.1	12.0	15.3	14.5	16.0	22.5	12.0	4.8			10.7	12.0	0.5	14.9
Garden Waste	5.2	9.0	13.8	12.5	9.0	--	12.0	9.6	69.7	69.0	10.4	35.7	64.1	16.3
Paper Products	60.3	50.0	42.4	42.5	48.0	45.5	42.0	56.5			40.6	30.0	30.2	34.9
Plastics/Rubber	2.0	3.0	1.8	4.0	2.0	2.5	2.4	--	2.2	0.2	4.6	--	1.3	6.4
Textiles	1.6	2.0	1.6	2.0	1.0	4.0	0.6	--	1.0	1.5	1.7	1.0	1.0	1.7
Wood	1.4	2.0	1.2	2.5	2.0	3.0	2.5	--	--	--	1.0	1.8	1.9	3.8
Metals	7.7	8.0	6.7	9.0	8.0	9.0	8.0	14.5	8.7	10.5	9.0	5.9	--	9.8
Glass/Ceramic	5.1	7.0	10.1	10.0	6.0	6.0	6.0	6.1	11.3	11.4	10.9	4.5	--	10.5
Ash/Dirt/Rock	6.6	7.0	7.2	--	8.0	7.5	11.0	8.5	--	--	2.8	--	1.0	1.7
Fines	--	--	--	--	--	--	3.0	--	--	--	8.3	7.0	--	--
Miscellaneous	--	--	--	3.0	--	--	0.5	--	7.1	7.4	--	2.1	--	--

[a] The number above each column is a reference number identifying the source of the data in the bibliography. In cases where a reference presents differing data (e.g., references 47 and 49), alternative waste streams have been analyzed.

[b] Average of all cells.

content of the various fractions of refuse is presented in Table 3 as a percentage of wet weight. As expected, food and garden wastes--the most rapidly decomposable fraction of refuse--contain the highest percentage of moisture, averaging over 50% moisture on a wet-weight basis.

Several studies have analyzed the chemical composition of typical refuse components; these data are summarized in Table 4. In cases where data for a particular chemical component are not shown, either the element was not detected in the refuse or it was not analyzed in that study. References are given for the data presented in Table 4.

Average values for the carbon, hydrogen, oxygen, nitrogen, and sulphur content of refuse components are given in Table 5.

The data in Tables 2 through 5 comprise the information base for estimating the theoretical maximum yield of methane from municipal refuse. The next chapter reviews two approaches to estimating methane production: (1) a stoichiometric method based on the chemical composition of the organic material to be biodegraded; and (2) a method based on the overall biodegradability of the refuse components.

Table 3. Moisture content of refuse components (for selected waste samples) (percent on wet weight basis).

REFUSE COMPONENT	MOISTURE CONTENT[a] 27[b]	47	48[c]	50
Food Waste	62	72	55	--
Garden Waste	--	65	47	32.7
Paper Products	20	10.2	24	24.6
Plastics/Rubber	20	2	16	--
Textiles	--	10	23	14.6
Wood	--	20	15	8.1
Metals	--	3	5	--
Glass/Ceramic	--	2	1	--
Ash/Dirt/Rock	--	10	14	--
Fines	--	--	32	--
Miscellaneous	--	4	--	7.4

[a] The number above each column identifies the source of the data in the bibliography.

[b] Mean values computed from all data given.

[c] Average of all cells.

Table 4. Chemical analysis of refuse components.

Reference[a]	47	47	47	47	47	47	47	47	47
ANALYSIS	Composite Refuse	Fats (Lipids)	Putrescibles	Metals	Glass Ceramics	Ashes	Paper	Wood	Grass
Percent Moisture (wet material)	20.7	0.0	72.0	3.0	2.0	10.0	10.2	20.0	65.0
Chemical Analysis (% dry material)									
C	28.0	76.7	45.0	0.8	0.6	28.0	43.0	50.5	43.4
H	3.5	12.1	6.4	0.04	0.03	0.5	5.8	6.0	6.0
O	22.4	11.2	28.8	0.2	0.1	0.8	44.3	42.4	41.7
N	0.33	0	3.3	--	--	--	0.3	0.2	2.2
S	0.16	0	0.52	--	--	0.5	0.2	0.05	0.05
Ash[b]	24.9	0.0	16.0	99.0	99.3	70.2	6.0	1.0	6.8
Volatiles[c]	75.1	100.0	84.0	1.0	0.7	29.8	94.0	99.0	93.2
Heat Content, kJ/kg	14 430	38 840	19 730	288	151	9704	17 610	20 040	17 890

Table 4. Continued.

Reference[a]	47	47	47	47	47	47	47	47	47
ANALYSIS	Brush	Greens	Leaves	Leather	Rubber	Plastics	Oils Paints	Linoleum	Textiles (rags)
Percent Moisture (wet material)	40.0	62.0	50.0	10.0	1.2	2.0	0.0	2.1	10.0
Chemical Analysis (% dry material)									
C	42.5	40.3	40.5	60.0	77.7	60.0	66.9	48.1	55.0
H	5.9	5.6	6.0	8.0	10.4	7.2	9.7	5.3	6.6
O	41.2	39.0	45.1	11.5	--	22.6	5.2	18.7	31.2
N	2.0	2.0	0.2	10.0	--	--	2.0	0.1	4.6
S	0.05	0.05	0.05	0.4	2.0	--	--	0.40	0.13
Ash[b]	8.3	13.0	8.2	10.0	10.0	10.2	16.3	27.4	2.1
Volatiles[c]	91.7	87.0	91.8	89.9	90.0	89.8	83.7	72.6	97.5
Heat Content, kJ/kg	18 380	16 460	16 510	20 590	26 350	33 420	31 190	19 330	17 800

Table 4. Continued.

Reference[a]	27	27	46	46	47	47	49	53
ANALYSIS	Paper	Garbage	Paper	Leaves	Street Sweepings	Dirt	Refuse	Compost
Percent Moisture (wet material)	--	--	4.07	1.04	20.0	3.2	--	--
Chemical Analysis (% dry material)								
C	44.71	43.19	51.21	55.72	34.7	20.6	39.0	35.7
H	7.44	9.40	5.85	5.37	4.8	2.6	--	--
O	40.66	26.64	41.14	28.72	35.2	4.0	--	--
N	0.41	2.90	0.04	4.20	0.1	0.5	0.56	1.07
S	1.28	1.14	0.12	0.11	0.2	0.01	--	--
Ash[b]	4.99	15.23	1.57	5.88	25.0	72.3	--	--
Volatiles[c]	95.01	84.77	98.43	94.22	75.0	27.7	--	--
Lipids	7.48	28.34	--	--	--	--	--	--

Table 4. Concluded.

Reference[a]	19	34	41	53
ANALYSIS	Refuse	Organics	Fines and Organics	Garden Waste
Chemical Analysis (% dry material)				
C	--	35.0	--	39.2
N	1.0	0.62	0.16	2.03
S	--	--	--	--
Ash[b]	54.0	--	88.0	21.0
Volatiles[c]	54.0	--	12.0	79.0
BOD_5, mg/L	--	--	0.6	--
COD,mg/L	--	--	3.6	--

[a] The reference number above each column refers to the source of the data.

[b] Noncombustible.

[c] 100 - % ash.

Table 5. Chemical composition of organic components in municipal refuse (percent on dry weight basis) (Ref. 51).

REFUSE COMPONENT	CARBON	HYDROGEN	OXYGEN	NITROGEN	SULPHUR	ASH
Food Waste	48	6.4	37.6	2.6	0.4	5.0
Garden Waste	48	6.0	38.0	3.4	0.3	4.5
Paper Products	44	6.0	44.3	0.3	0.2	5.5
Plastic	60	7.2	22.8	--	--	10
Rubber	78	10	--	2.0	2.0	10
Textiles	55	6.6	31.2	4.6	0.15	2.5
Wood	50	6.0	42.7	0.2	0.1	1.5
Ash/Dirt/Rock	26	3.0	2.0	0.5	0.2	68

ESTIMATION OF THEORETICAL MAXIMUM YIELD

ALTERNATIVE APPROACHES

Several different methods have been used to estimate the theoretical maximum yield of methane from municipal refuse. Each method necessarily assumes an efficiency of conversion and/or stoichiometry for the bioconversion of the organic matter to methane. The following section briefly reviews the approach of several investigators. Later sections of this chapter explore in more detail the methodologies employing (1) the basic stoichiometry of the conversion process, and (2) an approach utilizing an assumed conversion efficiency.

Using the stoichiometric method, the potential ultimate yield of methane gas has been estimated as 270 L CH_4/kg wet refuse. This estimate is based on a gross empirical formula representing the chemical composition of composite refuse ($C_{99}H_{149}O_{59}N$) and on the overall process stoichiometry (equation 1) approximating the combined mechanisms which take place during anaerobic decomposition of organic refuse in a landfill [24]. The same analysis estimated that the landfill gas would be 54% CH_4 and 46% CO_2 by volume.

A similar stoichiometric approach based on gross empirical formulas representing two groups of organic wastes constituting major sources of landfill gas, paper ($C_{203}H_{334}O_{138}N$), and food wastes ($C_{16}H_{27}O_8N$) resulted in an estimated ultimate yield of 230 L CH_4/kg wet refuse [24]. This approach also estimated an overall gas composition of 54% CH_4 and 46% CO_2.

An estimate of 415 L CH_4/kg refuse was reported by Dynatech R&D Company using the stoichiometric approach [56]. This estimate assumes that the refuse is cellulose on a dry weight basis and that all of it decomposes to methane. The stoichiometry for this reaction is:

$$C_6H_{10}O_5 + H_2O = 3CO_2 + 3CH_4$$

The estimated yield of methane would be 300 L CH_4/kg wet refuse.

The second method of estimating the ultimate gas yield is based on approximations of the overall biodegradability of "typical" composite refuse or of individual waste components. One approach assumed that 239 L of CH_4 could be produced from each pound of dry biodegradable volatile solids in the refuse [24]. Analyzing various refuse categories separately (based on the range of possible values for fractional composition, moisture and volatile solids content, and fraction of volatile solids biodegradable), and summing the gas contributions of the individual categories gave total methane yield estimates ranging from 6.2 to 230 L CH_4/kg wet refuse. Analysts using "average" characteristics for each refuse category gave a potential ultimate yield estimate of 47 L CH_4/kg wet composite refuse [24].

Another approach [43], based on the overall biodegradability of composite refuse, assumed the following: decomposable material constitutes 50% of in-place refuse weight; 50% of the decomposable material is volatile; 375 L of gas are produced for each kilogram of refuse destroyed [47]. Schwegler stated that the assumptions were conservative. These assumptions would result in an ultimate gas production estimate of 0.5 x 0.5 x 375 = 94 L gas/kg refuse. At a methane concentration of 50%, ultimate methane production would be 47 L CH_4/kg refuse.

Similar assumptions used elsewhere would give a higher yield estimate. For example, Pfeffer [42] assumed the following: 70% of domestic refuse is biodegradable; 70% of the biodegradable portion is anaerobically gasified; 687 L/kg of gas are formed for each pound of dry solids destroyed [44]. Further, assuming a 50% methane concentration and a 25% refuse moisture content would give an ultimate yield estimate of 0.7 x 0.7 x 0.5 x 0.75 x 687 = 125 L CH_4/kg wet composite refuse.

Numerous landfill gas recovery project reports have based ultimate yield estimates on the total organic carbon content of composite refuse [6, 8, 28]. This approach assumes the following: 1 mol of combined methane and carbon dioxide is produced from 1 mol of organic carbon; organic carbon content is 25 or 26% of the composite refuse by weight; 100% of the organic carbon is converted to gas. Estimates of ultimate yield based on these assumptions range from 506 to 530 L gas/kg composite refuse. Assuming a methane concentration in the gas of 50% by volume, the ultimate potential yield would be 260 to 270 L CH_4/kg composite refuse. The references differ as to whether these yields are based on wet or dry weight of composite refuse. However, the lower estimate of 506 L gas/kg of

Table 6. Summary of estimation methods for theoretical maximum methane yield.

Estimation Method	Estimated Yield L CH_4/kg wet composite	Assumptions Made	Reference
Balanced Stoichiometric Equations	230-270	Chemical composition of composite refuse, $C_{99}H_{149}O_{59}N$, and of paper ($C_{203}H_{334}O_{138}N$) and food wastes ($C_{16}H_{27}O_8N$).	24
Biodegradability of Materials	6.2-230 47 average	Assumes 1.5 kg biodegradable COD/kg volatile solids and 351 L/kg biodegradable COD	24
Biodegradability of Materials	47 average	Wet, composite refuse is 50% decomposable organics; 50% of decomposable organics is volatile; 375 L gas/kg volatile matter; 50% of gas is CH_4.	45
Biodegradability of Materials	120	Wet composite refuse is 70% decomposable organics; 70% decomposable organics converted to gas; 690 L gas/kg dry decomposable organics, 25% moisture content; 50% of gas is CH_4.	42
Total Organic Content	190-270	1 mol organic carbon yields 1 mol gas; CH_4 is 50% of gas produced, 100% of organic carbon is converted to gas.	5, 26, 4

refuse was stated to be on a dry weight basis. Assuming 50% CH_4 and 25% refuse moisture, this estimated yield could be as low as 187 L CH_4/kg wet composite refuse.

Table 6 summarizes the range of values for estimated ultimate methane yield for the different estimating methods described above.

From the above discussion, it can be seen that realistic, theoretical estimates of potential total methane production range from 47 to 270 L CH_4/kg wet composite refuse. We feel that, for conditions prevailing in most sanitary landfills, total methane production would fall within the range of 31 to 94 L CH_4/kg wet composite refuse. The total production can be enhanced by managing environmental factors favouring methane fermentation; in particular, moisture and pH.

Two stoichiometric methods used to evaluate the potential for methane production from organic solid wastes placed in landfills are examined below in detail. The first method utilizes the chemical composition of the refuse to compute on a strictly stoichiometric basis the maximum potential methane to be derived under optimal anaerobic conditions. The second method utilizes the gross composition of the refuse components along with an estimation of the biodegradability of the individual refuse components.

Tables 2 through 5 provide the data base for estimating the potential methane production by these two methods. Table 2 presents data on the percentage of various refuse components in solid wastes; Table 3, data on percent moisture content of refuse components; Table 4, values for the chemical composition of refuse components; and Table 5, data on the composition of organic components in municipal refuse.

STOICHIOMETRIC METHOD

The stoichiometric method of estimating potential methane production requires that the basic chemical composition of the organic material to be biodegraded be known. This method does <u>not</u> include any evaluation of the extent of biodegradability of the organic matter.

<u>Composite Refuse</u>

From Table 4, we will use the chemical composition of composite refuse as presented in a 1972 publication of *Solid Waste Management* [47].

Chemical composition of composite refuse:

	% (Wet Wt)	Normalized %	Mole Ratios	Stoichiometric Coefficients
C	28	51.6	4.3	99
H	3.5	6.5	6.5	149
O	22.4	41.3	2.58	59
N	0.33	0.6	0.0436	1
	54.23	100.0		

Based on the above computations, the empirical formula for refuse is:

$$C_{99}H_{149}O_{59}N$$

From equation (2) we can compute the fraction of substrate converted to cells, s

$$s = a_e \left[\frac{(1 + 0.2f\theta_c)}{(1 + f\theta_c)} \right]$$

Now, assuming the following values for the biological activity in the fill, $a_e = 0.2$, $f = 0.02$, and choosing an infinite time for θ_c, we get $s = 0.04$. We will use this value for general computation purposes; therefore, $e = 0.96$ (equation 1a).

The molecular weight of the empirical formula is:

$$M.Wt. = 12 \times 99 + 149 + 16 \times 59 + 14 = 2295$$

Utilizing the empirical formulation (equation 1), we have

$$d = 4n + a - 2b - 3c = 4 \times 99 + 149 - 2 \times 59 - 3 = 424$$

$$d = 424$$

$$\text{moles } CH_4 \text{ per mole composite refuse} = \frac{de}{8}$$

$$\frac{de}{8} = \frac{424 \times 0.96}{8} = 51$$

$$\text{moles } CO_2 \text{ per mole composite refuse} = (n - c - sd/5 - \frac{de}{8})$$

$$(n - c - \frac{sd}{5} - \frac{de}{8}) = 99 - 1 - \frac{0.04 \times 424}{5} - 51 = 44$$

Volume methane per unit weight of composite refuse:

$$\text{L } CH_4\text{/kg wet composite refuse} = \frac{(51 \times 22{,}266)}{(\frac{2295}{0.5423})}$$

$$= 268 \ \frac{\text{L } CH_4}{\text{kg wet composite refuse}}$$

where 1 kg-mol CH_4 occupies 22,266 L, at standard temperature and pressure.

The gas produced will be 54% CH_4 and 46% CO_2.

Refuse Components

An alternative approach to estimating potential methane production from refuse is to consider two groups of organic wastes which together constitute the major sources of landfill gas: (1) paper and garden wastes, and (2) food wastes. Using the data presented in Tables 4 and 5, stoichiometric coefficients for the constituent elements can be derived for these two organic waste categories. Based on the resulting empirical formulas and on assumed values for selected biological parameters, the volume of methane per kilogram of raw refuse in the two organic waste groups may be estimated.

Non-food Organic Wastes

Non-food organic wastes are principally comprised of garden wastes, paper, textiles, and wood. For purposes of calculating an empirical formula for this waste category, these refuse components are represented as paper materials.

Since paper is a wood-base product, we can look at the composition of wood for an evaluation of these materials. On a dry-weight basis, wood consists of: cellulose, 40 to 50%; lignin, 30%; hemicellulose, 15 to 25%; and other, 5%. Cellulose, a water-insoluble polysaccharide, constitutes the bulk of the cell membrane material of higher plants. It makes up 50% of wood and wood-like plant fibers, and over 90% of cotton fibers. Wood consists of molecules of cellulose, hemicellulose (a polymer of related structure), and lignin. Lignin, whose structure is not completely known, is a polymeric aromatic material which acts as a matrix, binding the cellulose fibers together. Cellulose (a polysaccharide made up primarily of glucose) and hemicellulose (a noncellulosic, nonpectin polysaccharide) are both substantially biodegradable under appropriate anaerobic conditions.

From Table 4 we obtain an average chemical composition of paper (average of three analyses) of:

	Normalized %	Mole Ratios	Stoichiometric Coefficients
C	48.84	4.07	203
H	6.69	6.69	334
O	44.20	2.76	138
N	0.26	0.02	1

These stoichiometric coefficients yield the empirical formula of $C_{203}H_{334}O_{138}N$ for paper.

Following the stoichiometric computation method used for composite refuse and again utilizing values of $a_e = 0.2$, $f = 0.02$, $s = 0.04$, and $e = 0.96$, we get the following results for potential methane from the garden waste, paper products, textile, and wood components of solid waste (here all computed as paper materials):

104 mol CH_4 per mole $C_{203}H_{334}O_{138}N$ (M.Wt. 4992 g/mol)

91 mol CO_2 per mole $C_{203}H_{334}O_{138}N$

yielding a gas composition of

53% CH_4
47% CO_2

and potential methane production of

466 L CH_4/kg Σx

where Σx = (garden waste + paper products + textiles + wood) on a dry weight basis. Potential methane production from Σx in raw refuse is estimated to be

209 L CH_4/kg raw refuse (wet weight).

This does not include the contribution from food wastes.

Food Wastes

The estimated contribution of potential methane production from the food waste component of refuse, computed by the same method as above, gives the following information (data base taken from Table 5):

Empirical Formula	= $C_{16}H_{27}O_8N$
Molecular Weight	= 361 g/mol
Gas Composition	= 60% CH_4 (8.64 mol CH_4)
	40% CO_2 (5.78 mol CO_2)
kg CH_4/kg food waste	= 535 L CH_4/kg food waste
kg CH_4/kg raw refuse	= 23.4 L CH_4/g raw refuse

Summary

The stoichiometric method of computing potential methane production yields the following:

	Material	L CH_4/kg Raw Refuse
1.	Composite Refuse	269
2.	Garden Wastes, Paper, Textiles, Wood	209
3.	Food Wastes	23.4
4.	(Sum of 2 + 3)	232.4

Presumably, the difference between methane production based on composite refuse (296) and the sum of the major organic fractions (232.4) is due to nonbiodegradable organics such as plastics, rubber, and other materials, as well as statistical variations in the data used for the basis of the computations. The values computed by the stoichiometric method assume optimal anaerobic conditions of moisture, pH, temperature, C/N ratio (weight of carbon divided by the weight of nitrogen), and micronutrients. Implicit in this approach is the absence of toxic or inhibitory substances which would impede the methane fermentation process. Obviously, these idealized conditions do not exist in most landfills. Nevertheless, the results obtained do put maximum boundaries upon the potential for methane production and will allow better focus upon those factors which might limit the achievement of optimal gas recovery from the waste materials.

METHOD UTILIZING BIODEGRADABILITY OF MATERIALS

This method of estimating potential methane production is based upon gross characteristics of the organic material, as characterized by volatile solids (V_i) and effective biodegradability (E_i) of the refuse components.

The following is a simple mathematical model for estimating methane production from refuse components based on data contained in Tables 2 to 4. Equation (4) estimates the maximum methane volume that can be produced from component i for a given weight (W_t) of wet bulk refuse when optimal conditions are available for the anaerobic conversion of organics to methane and carbon dioxide.

$$C = k \times k' \times W_t \times P_i \times (1 - M_i) \times V_i \times E_i) \qquad (4)$$

where,

W_t = Total weight of wet refuse
P_i = Fraction of component in total bulk refuse, wet weight basis
M_i = Fractional moisture content of component i, wet weight basis
V_i = Fractional volatile solids content of component c, dry weight basis
E_i = Fraction of dry volatile solids in component i that are biodegradable
C_i = Volume of methane gas from refuse component i (Potential methane production from unit weight of raw refuse is estimated by letting W_t = 1).

Coefficients/Constants

k = 351 L CH_4/kg biodegradable COD
k' = 1.5 kg COD/kg volatile solids

Total potential methane production from n refuse components is given by

$$C_t = \sum_{i=i}^{n} C_i \qquad (5)$$

Table 7 summarizes the typical range and averages for the various factors used in calculating the methane gas production per unit mass of refuse. Note that the data in this table do not correspond exactly to the numbers shown in Tables 2 through 4. Where appropriate, the authors have selected ranges and averages that represent, in their judgment, the most typical values.

Substituting in equation (4) the average values for the factors shown in Table 7, and letting W_t equal 1, one may calculate the methane gas production per unit mass for each organic component of the refuse. The results of these calculations are shown in the last column of Table 7. These results differ significantly from the potential methane production computed by the stoichiometric method (calculated as 230 to 267 L CH_4/kg wet composite refuse in the section entitled "Alternative Approaches"), differing by a factor of

Table 7. Calculation of theoretical maximum methane yield based on biodegradability of refuse composition[a].

Refuse Component (Organic Fraction)	Typical Range % In Raw Refuse [39]	Average % In Raw Refuse [39]	Typical Range % Moisture (wet wt.) [27,47,48,50]	Average % Moisture (wet wt.)	Average % Dry Solids (wet wt.)	% Volatile Solids (dry wt.) [27,46,47,49]	% Volatile Solids[b] Biodegrad. (dry wt.)	Methane Gas Production Per Unit Mass Refuse[c] Litres CH_4/kg
Food Waste	0.08-36	14.6	55-72	63	37	60	50	8.53
Garden Waste	0.3 -34.5	12.5	33-65	48	52	70	35	8.38
Paper Products	13 -61.8	42.7	10-25	20	80	85	20	30.57
Plastic/Rubber	0.3- 6.4	4.3	2-20	13	87	95	2	0.37
Textiles	0.1- 8.9	2.4	10-23	16	84	94	5	0.50
Wood	0- 7.5	2.5	8-20	14	86	85	5	0.48

[a] Note that the data on this table do not correspond exactly to the numbers shown in Tables 2 through 4. Where appropriate, the authors have selected ranges and averages that represent, in their judgment, the most typical values. The numbers under the column titles identify the sources of the data in the bibliography.

[b] Estimated.

[c] 47.5 L/kg.

about 5. It is reasonable to expect considerable variation in composition of refuse and in landfill environments, and these estimates are based on two different methods, giving only approximate results. No one value for potential methane production is merited.

Again, it must be emphasized that all the above estimates are for theoretical maximum CH_4 production under optimal anaerobic conditions which are seldom obtained throughout the fill volume. On the other hand, under non-optimal anaerobic conditions, and assuming less than 100% efficiency for recovery of methane, a figure of 60 ± 30 L of CH_4/kg of raw refuse represents a practical estimate of actual recovery potential.

COMPARISON OF OBSERVED AND CALCULATED GAS COMPOSITION

The composition of landfill gas produced under anaerobic conditions is typically observed to be 50 to 70% methane and 30 to 50% carbon dioxide. Traces of ammonia, hydrogen sulphide, and other gases may also be present. Quantities of oxygen and nitrogen encountered in analysis typically result from air contamination during sampling, from atmospheric air that may be drawn through the fill when an induced exhaust gas extraction system is employed, or from air leaks in the gas collection system.

Table 8 presents typical landfill gas composition data observed at a number of landfills. As indicated in footnote a of this table, the calculated gas composition using the stoichiometric method (described earlier in this report) compares quite favorably with field measurements of gas composition.

Because of its combustible nature, methane is the gas of principal interest when gas hazard control or gas recovery is contemplated although other potentially corrosive gases may be of concern. Methane, a colourless, odourless hydrocarbon, is combustible at concentrations of 5 to 12% by volume in air and has a heating value of approximately 37,260 kJ/m^3 of methane. Since methane averages only about 55% of the landfill gas composition, the heat content of the gas as produced in the landfill is about 20,490 kJ/m^3. During recovery or harvesting of the gas, some reduction in methane concentration occurs due to dilution with air, and the more usual value of heat content for the extracted gas is 16,770 to 18,630 kJ/m^3.

Table 8. Typical landfill gas composition[a].

LANDFILL	Methane %	Carbon Dioxide %	Nitrogen %	Oxygen %	Other %
Azuza Western Azuza, CA	50	50	--	--	--
Bradley Los Angeles, CA	50	50	--	--	--
Central Disposal Site Sonoma County, CA	50	50	--	--	--
G.R.O.W.S. Morristown, PA	46	53	1	--	--
Hewitt Los Angeles, CA	45	55	--	--	--
Mountain View[b] Mountain View, CA	44	34	21	1	--
Palos Verdes[b] Rolling Hills, CA	53	43	3	--	1
P.I.I. Denver, CO	45	55	--	--	--
Scholl Canyon[b] Glendale, CA	40	51	7	2	--
Sheldon Arleta Los Angeles, CA	55	45	--	--	--

[a] The data presented in this table were obtained from an unpublished study by EMCON Associates; they are based on field measurement of gas composition, and are within the range of the theoretical values computed by the stoichiometric methods described earlier in this chapter. The computed values for methane and carbon dioxide are as follows: composite refuse, 54% and 46%; paper products, 53% and 47%; food wastes, 60% and 40%, respectively.

[b] The presence of oxygen and nitrogen is believed to be caused by the air drawn into the fill during the extraction process.

TIME DEPENDENCY OF GAS PRODUCTION

GENERAL CONSIDERATIONS

During the relatively early active life of a landfill, the composition of landfill gas undergoes an evolutionary process as the waste experiences first aerobic, and then anaerobic environments. The changes in gas composition can be characterized by four relatively distinct phases (Figure 4). In the first phase, lasting several days to weeks, oxygen is present as a component of air from the time of waste placement; carbon dioxide is the principal gas produced during this stage. During the second phase, anaerobic conditions prevail once oxygen has been depleted. As anaerobic decomposition begins, significant amounts of carbon dioxide and some hydrogen are produced. The third phase, also anaerobic, is characterized by the first evidence of methane, a reduction in the amount of carbon dioxide produced, and the depletion of hydrogen. The fourth phase is also anaerobic and differs from the third in that gas production and composition approach pseudo-steady-state conditions.

The amount of time associated with each phase of gas production is a function of the specific conditions within a given landfill. Once methane production begins within a landfill, it usually will continue over a number of years, the total time depending on site-specific conditions. Gas generation may range from a few years to several decades in certain environments.

IMPORTANT SYSTEM VARIABLES

The rate of gas production at any time is a function of numerous factors, including: (1) size and composition of refuse; (2) age of the refuse; (3) moisture content of refuse; (4) temperature conditions in landfill; (5) quantity and

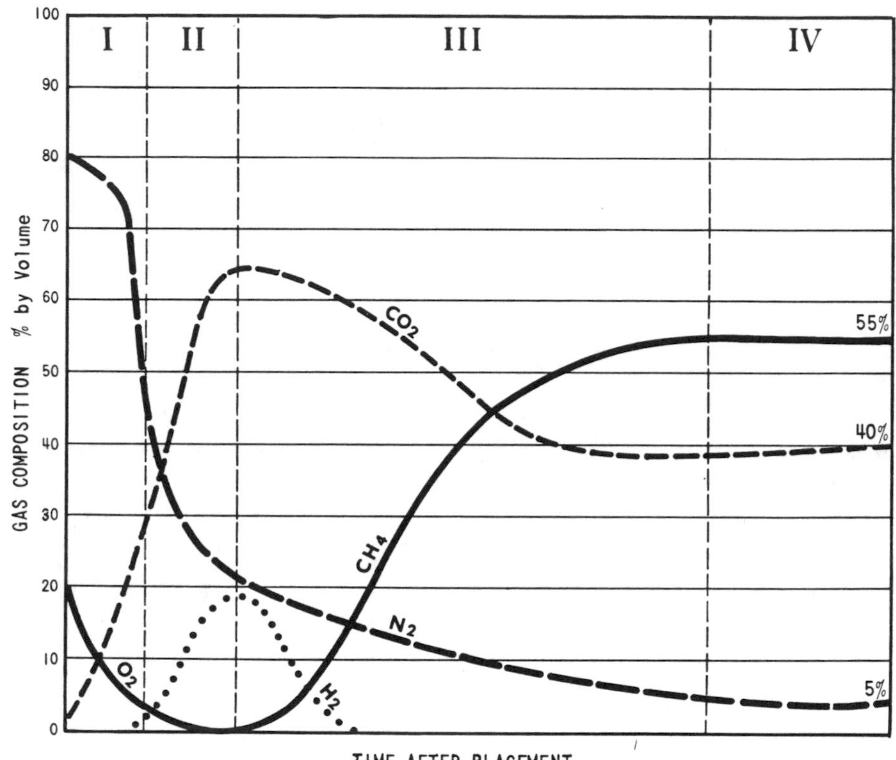

I. Aerobic

II. Anaerobic, Non-Methanogenic

III. Anaerobic, Methanogenic, Unsteady

IV. Anaerobic, Methanogenic, Steady

Figure 4. Evolution of typical landfill gas composition.

quality of nutrients; and (6) pH and alkalinity of liquids within the landfill (probably only important for those landfills actively producing leachate). As the landfill ages, gas production gradually decreases, and the landfill eventually returns to an aerobic environment when the biodegradable materials are depleted.

Several of the landfill variables are potentially manageable by the landfill operator/manager. These variables could be manipulated to enhance both the rate and quantity of methane produced within the landfill. For instance, the following factors could potentially be altered or managed: (1) composition of refuse materials, (2) exclusion of toxic or inhibitory materials, (3) time/space of refuse placement within a landfill volume, (4) moisture content and moisture recycling, (5) nutrient addition, (6) pH control, (7) particle size of refuse materials (size reduction), (8) density of in-place refuse (reflects degree of compaction), (9) permeability and porosity, (10) microbial population (seeding). Temperature is probably the one parameter which is not readily amenable to management. Current practice commonly does not allow for management of any of the above variables.

Factors Affecting Production Rate

Refuse composition directly affects the rate of methane production, as well as the ultimate yield. Maximum methane production is stimulated by a waste having a high percentage of biodegradable materials (food and garden wastes, paper, textiles, and wood). Composition changes reflect seasonal and geographical variations; e.g., significant increase in garden trimmings in fall, seasonal agricultural wastes, higher percentage of garden wastes in tropical or fast-growing geographical areas, sludge applications, etc.

Studies of anaerobic waste treatment in digesters indicate that no toxic or inhibitory materials should be present in the waste since such substances readily upset the activity of methanogenic bacteria. Industrial wastes may contain inhibitory concentrations of common salts of sodium, potassium, magnesium, calcium, ammonium or sulphide, and, more importantly, toxic organic solvents such as carbon tetrachloride and chloroform, among others [28]. During anaerobic digestion of sewage sludge, it has been observed that, while sodium, potassium, calcium and magnesium may stimulate gas production at relatively low concentrations (75 to 400 mg/L), inhibition results from higher concentrations (1000 mg/L), and toxic effects are noted at still higher concentrations (e.g., 2000 to 3000 mg/L for calcium) [28].

As already noted, methane production begins only after all oxygen has been depleted; indeed, oxygen is toxic to the strictly anaerobic methanogenic bacteria. If a gas recovery project is implemented, high gas extraction rates may cause problems by creating a pressure gradient across the refuse cover or perimeter, drawing in atmospheric oxygen and thereby "poisoning" portions of the methane-generating landfill. Overstressing of the system must be avoided if an optimal ongoing gas recovery project is to remain economically viable. Ideally, a large number of closely spaced recovery wells would be utilized, and the extraction rate of each well would be relatively low. The extraction rate could be closely regulated by oxygen-sensing, or pressure-sensing instrumentation. Of course, there is usually an economic trade-off limiting the initial capital which may be expended during system installation, and resulting in less than ideal recovery system efficiency.

A high moisture content, in the range of 60 to 80% (wet weight basis), favours maximum methane production rate. Studies have shown, for example, that gas production can increase after a heavy rainfall. In reality the moisture content of refuse at the time of placement is normally well below this range, averaging about 25%. In an effort to control leachate production, landfill design and operation typically focus on limiting the refuse moisture content. Consequently, by environmental design, lack of moisture may significantly limit gas production rate in a sanitary landfill.

Although methane formation will proceed in a pH range from 6.5 to 8.0, the optimal pH for methane fermentation is in the neutral to slightly alkaline range, between 7.0 and 7.2. A drop of pH below a value of about 6 may be toxic to methanogenic bacteria. Most landfills have an acidic environment initially but, within the first several years, the pH rises toward neutrality. Therefore, optimal pH conditions for maximum methane production typically do not occur for several years. As noted in Figure 3, the presence of significant alkalinity will tend to buffer the system, minimizing sudden changes in pH and stabilizing the gas-production process.

DeWalle has shown that refuse size reduction can have a marked effect on the rate of gas production, presumably because of the increased surface area available for organisms to attack the organic materials [11]. In addition, data indicate that increased moisture content can have a direct and proportionate effect on the rate of gas production [11].

Warm landfill temperatures favour methane production; a dramatic drop in activity has been noted at temperatures below 10°C. Production rates for methane may thus reflect seasonal temperature fluctuations in cold climates where the landfill is shallow and responsive to ambient temperatures. Gas temperatures as high as 71°C have been observed in a deep landfill (average refuse thickness of 37 m. Optimum temperatures for anaerobic digestion of sewage sludge have been reported as 29 to 38°C for mesophilic operation and 49 to 57°C for thermophilic operation [28].

Optimum anaerobic conditions for rapid gas production are rarely, if ever, observed in normally operated landfills. The rate of gas production may be limited by any of the conditions discussed above. In addition, mass transport is probably a rate-limiting factor in the typical landfill since the contact opportunity between the organisms and the organic substrates or inorganic nutrients is very limited. Control of temperature in the landfill may be impractical, but the other factors affecting the rate of gas production might be controlled to a varying extent.

Enhancement of Gas Production

Means of increasing methane production include changing the composition mix of the refuse by (1) increasing the organic content, and (2) by limiting the presence of toxic or inhibitory substances. The organic content may be increased by the addition of sewage sludge, manure or agricultural wastes; removal of ferrous and nonferrous metals; separation of heavy and light material; and use of less daily and intermediate cover soil. The types and amounts of waste admitted to the landfill can be carefully screened to avoid placement of substances in toxic or inhibitory concentrations.

Anaerobic conditions can be maintained by careful operation of the gas recovery system, thereby avoiding air intrusion that can be caused by excessive gas extraction rates. In addition, the surface and perimeter of the landfill can be sealed with clays, synthetic materials, or paving to prevent air intrusion (the sides and/or bottom could be sealed during fill construction).

Refuse moisture content could be increased during placement by the addition of water or sludge (a dual benefit in that organic content could also be increased). After refuse placement, moisture content could be increased by infiltration into the fill of water applied at the surface.

Environmental pH, alkalinity, and nutrient availability can be improved by the addition of chemicals during refuse placement. Leachate recirculation would permit subsequent control of these conditions and would improve mass transport within the landfill.

The practical feasibility of gas enhancement appears excellent. Pacific Gas and Electric Company (PG&E) has appropriated funds for an extensive gas enhancement program at the Mountain View Landfill in California to supplement their current gas recovery project. Dynatech, Easley and Brassey, and EMCON Associates have been selected to assist PG&E in this innovative program which will involve size reduction, sludge addition, and pH control.

OBSERVED PRODUCTION RATES

Estimates of the rate of gas production from operating, as well as from recently closed landfills, range from a low of 1.3 L CH_4/kg/yr of wet refuse, to a high of 7.5 L CH_4/kg/yr. For comparative purposes, Table 9 presents data for landfills where recovery projects are either on-going or under study.

Within a few years of refuse placement, production rates on the order of 2.5 to 3.7 L CH_4/kg/yr of wet refuse commonly are reported. Laboratory characterization of the landfilled refuse in terms of total organic carbon, volatile solids, and moisture will give a preliminary indication of whether production rates above or below those typically observed should be anticipated. Normally observed refuse characteristics might be 25% for moisture (wet weight basis) and total organic carbon, and 50 to 75% for volatile solids.

THEORETICAL KINETIC MODELS FOR GAS PRODUCTION

At the present time there are no available field data from sanitary landfills which will allow verification of a kinetic model to describe the time dependency of gas production. There are, however, models describing the growth kinetics of bacterial populations, and there is no reason to suspect that the heterotrophic anaerobic organisms in sanitary landfills should deviate from these general models.

Although the importance of many physical and chemical variables is known for the anaerobic degradation processes, it is not possible to describe with any certainty the actual conditions inside a landfill system. In addition, no explicit functional relationships exist between many of the factors known to affect the anaerobic microbial systems (e.g., moisture content) and the relationships expressing the kinetics of anaerobic gas production. Therefore, the approach currently taken in modelling efforts is to use the most simplified model available consistent with fundamental principles, and to empirically adjust the kinetic rate constants to account for variations in such variables as moisture content and temperature. The

Table 9. Landfill gas recovery comparison data[a].

Landfill Name	Year Fill Began	Year Fill Completed	Refuse In Place kg x 10^9	Surface Area ha	Average Thickness of Refuse m	Predicted Methane Extraction Rate L/sec	Annual Methane Production per Kilogram of Refuse L/kg/yr
Azuza Western Azuza, CA	1953	Still Filling	6.0	22	37	425	2.5
Bradley Sun Valley, CA	1960	Still Filling	7.5	24	37	613	2.5
Coyote Canyon Irvine, CA	1964	1981 (est)	19.6	162	NA	1557	2.5
Hewitt Los Angeles, CA	1962	1975	5.6	24	31	472	2.5
Mountain View Mountain View, CA	1975±	1975±	0.7	8	12	165	7.5
Palos Verdes Rolling Hills Estates, CA	1957	1975	3.4	13	31	330	3.1
Scholl Canyon Glendale, CA	1963	1974	4.3	18	27	189	1.2
Sheldon Arleta Los Angeles, CA	1962	1974	2.7	15	26	472	5.6

[a] Data obtained from an unpublished study by EMCON Associates.

[b] Figures based on published or latest available data, including personal communication.

simplified kinetic model for the production of CH_4 and CO_2 described below cannot account explicitly for most of the known important variables. Variations in physical-chemical environments would be accounted for in the overall rate constant.

The kinetic expression used here to describe the time rate of production of CO_2 and CH_4 is a pseudo first-order equation. The pseudo first-order expression results from applying limiting conditions to the classical Monod equation relating the rate of substrate utilization both to the concentration of microorganisms in the system and to the concentration of soluble substrate surrounding the organisms [37]. This equation, in only a slightly modified form from that used by Monod [37], is

$$\frac{dS}{dt} = \frac{KXS}{K_s + S} \tag{6}$$

in which K is the maximum rate of substrate utilization per unit mass of microorganism (occurring at high substrate concentration) [$time^{-1}$]; X is the concentration of microorganisms [mass/volume]; S is the concentration of substrate surrounding the microorganisms [mass/volume]; and K is the waste concentration at which the rate is one-half the maximum rate of substrate utilization [mass/volume]. Equation (6) indicates that the functional relationship between substrate utilization rate and substrate concentration is continuous over the total range of substrate concentrations. In the two extreme cases, when S is very large ($S \gg K_S$), and when S is very small ($S \ll K_S$), equation (6) can be approximated by the following discontinuous functions:

$$\frac{dS}{dt} = KX, \text{ when } S \gg K_s \tag{7}$$

$$\frac{dS}{dt} = \frac{K}{K_s} X \quad S, \text{ when } S \ll K_s \tag{8}$$

Equation (7) is zero order with respect to substrate concentration and equation (8) is first order with respect to substrate concentration. A necessary assumption in applying this kinetic model to the decomposition of organics in a sanitary landfill is that the organic waste is the limiting nutrient for the rate-determining methane bacteria [33]. A recent study by Chan and Pearson [8] on the hydrolysis of cellulose indicated that hydrolysis of insoluble cellulose to soluble cellobiose appeared to be the rate-limiting step in the anaerobic decomposition of cellulose. They also concluded that the hydrolysis of cellulose to cellobiose is not only the limiting step of the cellulose hydrolysis process, but is also the rate limiting step in the overall cellulose fermentation process. Since a major fraction (≈70%) of organic material in refuse is cellulose based, this supports both the notion of substrate limitation on kinetics and the critical role moisture content plays in the degradation process. The rate-determining nature of methanogenic bacteria and the effects of environmental and metabolic factors on the kinetics of methane fermentation have been studied in great detail for applications to anaerobic digesters [28, 33]. Little of this information can be applied directly to organic decomposition in sanitary landfills because of the undefined nature of the landfill environment [25].

The three case studies presented below all have first-order kinetics as part of the overall kinetic model. However, there are significant differences in approach and assumptions, as will become apparent. It is important to note that, although these kinetic models are useful as references, they have not been verified by field data.

Palos Verdes Kinetic Model

In the Palos Verdes report, a two-stage, first-order mathematical model was employed to represent the kinetics of gas production in a landfill. It was assumed that the first-stage gas production rate (dG/dt) is proportional to the volume[1] of gas already produced (i.e., gas production rate is increasing exponentially with time). During the second stage of gas production, it was assumed that the rate of decrease of the remaining potential for gas production (-dL/dt) is proportional to the volume of gas remaining to be produced (i.e., gas production rate is an inverse exponential function of time). This two-stage model can be described mathematically, as follows:

First Stage:

$$\frac{dG}{dt} = K_1 G \tag{9}$$

Second Stage:

$$\frac{dL}{dt} = -K_2 L \tag{10}$$

where,

t = time

G = volume of gas produced prior to time t

L = volume of gas remaining to be produced after time t

---

[1] Volume of gas is being used here as a measurement of gas quantity and would have to be consistently determined relative to the same temperature and pressure.

k_1 = first stage gas production rate constant

k_2 = second stage gas production rate constant

It was further assumed that the maximum gas production rate and the transition from first-stage to second-stage kinetics occurs at the time when half of the ultimate gas production has been reached (i.e., $G = L = L_o/2$ when $t = t_{1/2}$), referred to here as the "half-time". Two ways of expressing the limits of integration, and relationships resulting from integration of the first-stage equation follow:

$G = G_o$ when $t = 0$ or $G = G$ when $t = t$

$G = G$ when $t = t$ $\quad$ $G = L_o/2$ when $t = t_{1/2}$

for $t \leq t_{1/2}$

$$\ln G = \ln G_o + k_1 t \qquad (11)$$

or

$$G = G_o e^{k_1 t} \qquad (12)$$

Since $G = L_o/2$ when $t = t_{1/2}$,

$$\ln G = \ln \left(\frac{L_o}{2}\right) - k_1 \; (t_{1/2} - t) \qquad (13)$$

or

$$G = \frac{L_o}{2} \; e^{-k_1 \; (t_{1/2} - t)} \qquad (14)$$

Limits of integration for the second stage equation and the relationships resulting from the integration are as follows:

$L = G = L_o/2$ when $t = t_{1/2}$

$L = L_o - G$ when $t = t$

for $t^{1/2} \leq t$

$$\ln L = \ln (L_o/2) - k_2 \; (t - t_{1/2}) \qquad (15)$$

$$L = (L_o/2) \; e^{-k_2(t-t_{1/2})} \qquad (16)$$

$$G = L_o - L = L_o \; [1 - 1/2 \; e^{-k_2(t-t_{1/2})}] \qquad (17)$$

In the Palos Verdes report, the organic wastes were considered in three main categories: readily decomposable wastes (food and grass); moderately decomposable wastes (paper, wood, and textiles); and refractory wastes (plastic and rubber). For purposes of estimating values of the first-stage gas production rate constant (k_1), a value of $t_{1/2}$ was assumed for each of the three categories of organic refuse. It was also assumed that the value of G_o is $L_o/100$ for each waste category. A value of k_1 can then be calculated by rearranging equation (11) to give the following expression:

$$k_1 = \frac{\ln\left(\frac{G}{G_o}\right)}{t}$$

Substituting in the assumed value of $G_o = L_o/100$ and the value $G = L_o/2$ at the assumed half time, $t_{1/2}$, permits calculation of k_1 as follows:

$$k_1 = \frac{\ln(50)}{t_{1/2}} \tag{18}$$

For purposes of calculating k_2, values of $t_{99/100}$ (the time required to achieve 99% of the ultimate gas production), were assumed for each category of organic refuse. Rearranging equation (15) gives the following:

$$k_2 = \frac{\ln\left(\frac{L_{o/2}}{L}\right)}{t - t_{1/2}}$$

Substituting $L_o/100$ for L at $t = t_{99/100}$ yields the following relationship, which can be used to calculate k_2:

$$k_2 = \frac{\ln(50)}{t_{99/100} - t_{1/2}} \tag{19}$$

Apparently, the values of k_2 tabulated in the Palos Verdes report were miscalculated, as those k_2's equal

$$\frac{\ln(100)}{t_{99/100} - t_{1/2}}$$

From equations (18) and (19) it can be seen that calculated values of k_1 and k_2 do not depend on the value assumed for L_o (at least, when G_o is assumed to be some fraction of L_o). It is the assumed values of $t_{1/2}$, $t_{99/100}$ and G_o which establish the values of the rate constants. Of course, values of k_1, k_2 and G_o could be initially assumed, permitting calculation of $t_{1/2}$, $t_{99/100}$, gas production at a given time, and time required to reach a given gas production. Table 10 gives values assumed in the Palos Verdes report for $t_{1/2}$, $t_{99/100}$ and refuse composition.

Table 10. Assumed values for selected kinetic factors (Palos Verdes study).

	Assumed Percentage of Total Organic Fraction[a] (wet basis)	Assumed $t_{1/2}$ (years)	Assumed $t_{99/100}$ (years)	Calculated k_1 (years^{-1})	Calculated k_2 (years^{-1})	Reported k_2 (years^{-1})
Readily Decomposable Organics	35.4	1	3.5	3.91	1.56	1.84
Moderately Decomposable Organics	61.0	2	6	1.96	0.978	1.15
Refractory Organics	3.6	20	60	0.196	0.0978	0.115

[a] The percent composition is based on data from "Municipal Refuse Disposal", Institute for Solid Wastes, American Public Works Association, 1970. These percentages do not necessarily reflect the composition of refuse at the Palos Verdes Landfill.

Assumed and calculated values of L_o,G_o, $t_{1/2}$, k_1, and k_2 can be substituted into equations (12) or (14), and (17), which can then be used to calculate an estimate of gas volume produced from a waste category prior to any time, t. Production rate at any time can be estimated from the derivatives with respect to time of these equations. Total gas production volume is the sum of the productions from each refuse category. For any assumed total ultimate yield, L_o, the Palos Verdes report assumed that the ultimate yield from each category of waste, L_{oi}, is equal to $P_i x L_o$, where P_i is component i's percentage of the total organic fraction of the refuse, on a wet weight basis.

Figure 5 graphically depicts methane gas production, as estimated by the Palos Verdes kinetic model. The upper graph presents the log of the rate of gas production over time for each of the three refuse components. Note that each of the three peaks has a positive slope on the left of the maximum, and a negative on the right; the positive slope corresponds to the first stage of methane production and the negative to the second stage. These schematic curves are based on the assumption that the readily decomposables provide 20% of the total gas production; the moderately decomposable, 50%; and the refractory, 30%. Note that these percentages differ from those presented in Table 10.

The lower graph in Figure 5, which shows the fraction of gas produced per unit mass of refuse, is based on the data presented in the upper graph; the composite curve shown in the lower graph is the summation of the three components depicted in the upper graph.

In employing the Palos Verdes kinetic model, it should be noted that the first-stage equation is not suitable for application at the inception of methane formation in a landfill. The gas production rate is said to be proportional to the volume of gas already produced, G. Since G=0, at a landfill's inception, production rate would remain zero. As stated above, the Palos Verdes report assumed that the first-stage equation became applicable when gas production reached 1% of the ultimate yield (i.e., when G_o=L_o/100).

In the Palos Verdes report, it has also been noted that ultimate gas production for each category of organic refuse was computed as the product of that component's fractional weight of the total organic content on a wet basis times the assumed total, ultimate gas production. Perhaps each component's ultimate production would better have been estimated on the basis of that component's fraction of the total organic content on a dry basis and the average carbon content of each component.

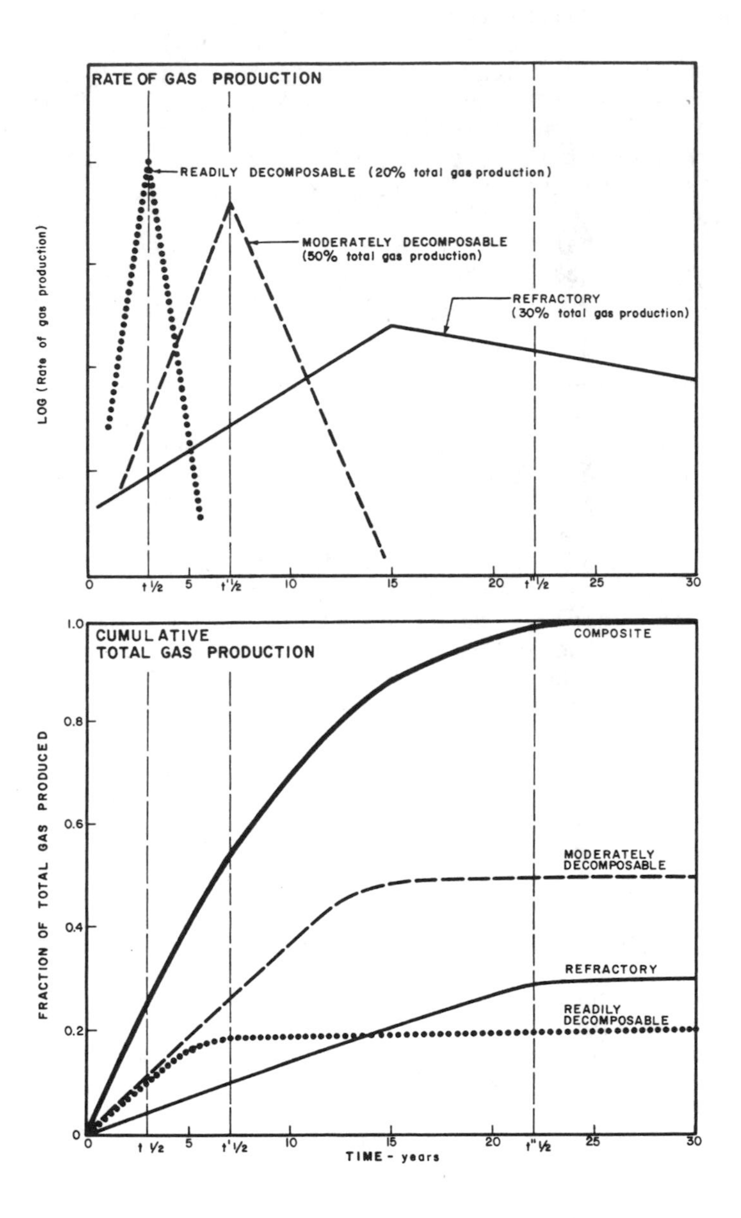

Figure 5. Rate of gas production and cumulative total gas production (Palos Verdes Kinetic Model).

Finally, in the Palos Verdes report, the maximum production rate and the rate constants (therefore, the extent of the gas production curves along the time axis) were established by (1) the assumption that the point of inflection in the gas production curves occurs when $t=t_{1/2}$, (2) the assumption that $G = G_0=L_0/100$, and (3) the assumed values for $t_{1/2}$, $t_{99/100}$, and L_0. The first two assumptions have already been addressed. It should also be noted that the values of $t_{1/2}$ and $t_{99/100}$ may have been considerably underestimated. The economical gas production life of a "typical" landfill is probably significantly greater than the six years mentioned in the Palos Verdes report.

Sheldon Arleta Kinetic Model

Since gas recovery work at the Sheldon Arleta Landfill utilized a two-stage, first-order kinetic model similar to that employed at Palos Verdes, many of the above comments and criticisms apply. The approach in the Sheldon Arleta report was based on an article by G.M. Fair and E.W. Moore related to anaerobic digestion of sewage sludge and on a master's thesis by Robert Alpern concerning decomposition rates of landfilled refuse [13, 1]. The Fair and Moore article presented a plot of gas production versus time for sewage sludge digestion, and the Sheldon Arleta estimation of the time dependency of gas production was based on the Fair and Moore curve for total gas production, including both methane and carbon dioxide. The Fair and Moore curve showed the peak digester gas production rate occurring at 14 days, and gas production being about 99% complete at 40 days [13]. For the Sheldon Arleta report, this curve was normalized along both axes; hence, the approach assumes that maximum landfill gas production rate occurs at 14/40=0.35 of the time required for 99% complete gasification. It was further assumed that the time required to reach the maximum production rate coincides with the "half-time", $t_{1/2}$ (i.e., $t_{1/2}=0.35t_{99/100}$). The dimensionless gas production curve, as generated from the Fair and Moore curve, was used to estimate the time dependency of gas production for incremental additions of refuse to the landfill; the incremental productions were then summed to give total production.

The Sheldon Arleta landfill was operated over a 12-year period, and the refuse weight landfilled each year was considered separately in the analysis of gas production kinetics. In addition, the refuse landfilled each year was considered in two separate categories; readily decomposable materials (e.g., garbage, grass, and tree trimmings) were distinguished from relatively slowly decomposing materials (e.g., newspaper, cardboard, and lumber). In all, the total refuse mass was divided into 24 parts which were considered separately and summed for total gas production. The following assumptions were utilized to estimate the maximum theoretical gas yield for each of the 24 refuse subdivisions:

- composite refuse is 26% carbon by weight
- approximately 31% of the refuse carbon is readily decomposable
- approximately 66% of the refuse carbon is more slowly decomposable
- 100% of the carbon is converted to gas
- one kilogram of carbon yields 2.01×10^3 L of gas.

Half-times were assumed for each of the two refuse categories, and the time required for total (i.e., about 99%) gasification was estimated from the observation that on the Fair and Moore curve and its dimensionless equivalent, $t_{total} = t_{1/2}/0.35$. For each refuse category this value of t_{total} and the dimensionless curve of production rate permitted estimation of the ratio of production rate at any time to the maximum production rate; this ratio was estimated for each year up to t_{total}. Because of the assumption that the half-time, $t_{1/2}$, coincides with the time of maximum production rate, values of the production rate ratio were modified to give equal areas under the dimensionless curve on either side of a vertical line drawn at the time of maximum gas production rate (now equal to $t_{1/2}$). The area under the entire curve represents total, maximum, theoretical gas production, L_0; the area under the curve on either side of $t_{1/2}$ represents a gas production volume of $L_0/2$. For each refuse category, average annual production rates for each year were then calculated from the theoretical ultimate gas production volume (L_0) and the ratio of area under the modified curve for a given year to the total area under the modified curve. In this way, average annual production rates from the time of refuse placement to t_{total} were estimated for each of the 24 refuse subdivisions. Average annual production rates for the entire refuse mass for each year were calculated as the sum of the rates of the 24 refuse subdivisions during that year.

The Sheldon Arleta approach was performed with various initial assumptions of half-times for the two refuse categories, until the calculated total production rate at a point in time matched a prior, independent estimate of production rate at that same point in time. The half-times and total production times ($t_{1/2}/0.35$) finally accepted in the report are as follows:

	$t_{1/2}$	t_{total}
Readily Decomposable	9 yr	26 yr
More Slowly Decomposable	36 yr	103 yr

Comparison with the values assumed in the Palos Verdes report shows that the estimates differ by factors ranging from 7 to 18.

Since the Sheldon Arleta model is almost identical to the Palos Verdes model, its graphs are similar to those shown in Figure 5. The main difference is that for Sheldon Arleta the log of the rate of gas production has two peaks instead of the three shown in Figure 5. The cumulative gas production curve for Sheldon Arleta would be similar to that shown in Figure 5.

In applying the Fair and Moore curve (gas production versus time for the anaerobic digestion of sewage sludge) to gas production in a landfill, the Sheldon Arleta approach explicitly assumed that the gas production rate in a landfill is not limited by lack of moisture or nutrients. Nevertheless, the report did acknowledge that "moisture is critical to the anaerobic digestion process..." and that "...very little moisture is added to sanitary landfills..." [26]. Implicitly, the approach assumed that the possibility of rate limitation by inadequate mass transport, unfavourable pH, or oxygen intrusion did not preclude adapting the sewage sludge digester model to the landfill environment. These considerations were addressed in the discussion of the Palos Verdes kinetic model and will not be expounded upon here.

As in the Palos Verdes approach, the Sheldon Arleta report assumes that the maximum production rate occurs at the time when half of the ultimate gas production has been attained (i.e., when $t=t_{1/2}$) [26]. It has already been noted that, although this assumption may be approximately true for anaerobic digestion of sewage sludge, maximum gas production rate from a landfill probably occurs at a time substantially shorter than $t_{1/2}$. In a sewage sludge digester under optimal conditions, the microbial mass may expand geometrically until limited by a decrease in the substrate (organic waste) concentration below some critical level; however in a landfill, many factors may limit biomass growth before the total amount of remaining organic matter becomes limiting.

It is interesting to note that, even for the normalized Fair and Moore curve, time of maximum gas production rate (t_{max}) occurs prior to $t_{1/2}$. About 40% of the total area under the curve lies to the left of t_{max}, indicating that t_{max} equals $t_{0.4}$ rather than $t_{0.5}$ for this model of anaerobic sludge digestion. The Sheldon Arleta approach manipulated coordinates of the curve along the ordinate so that $t_{0.5}$ was made to equal t_{max}. As a result, the Sheldon Arleta approach estimates that 50% of the ultimate yield is attained by the time gas production rate reaches its maximum, followed by a rapid decrease in production rate. By comparison, the original sludge digester model estimates that 40% of the ultimate yield is attained by the time of maximum production rate, followed by a less rapid (relatively speaking) decrease in production rate.

In reality, maximum production rate in a landfill would probably occur when substantially less than even 40% of the ultimate yield had been attained, followed by a relatively slow decrease in production rate. According to the Sheldon Arleta kinetic model, it would take more than 30 years for a landfill to reach its maximum gas production rate.

Scholl Canyon Kinetic Model

The Scholl Canyon report adopted a single-stage, first-order kinetic model similar to the second-stage model of the Palos Verdes and Sheldon Arleta approaches [12]. The Scholl Canyon approach assumes that, after a lag time of negligible duration, during which anaerobic conditions are established and the microbial biomass is built up and stabilized, the gas production rate is at its peak. Thereafter, the gas production rate is assumed to decrease as the organic fraction of the landfilled refuse (measured as remaining methane production potential, L) diminishes and the landfill is able to support an ever decreasing biomass of gas-producing microorganisms. This is a model of substrate-limited microbial growth, as described by the following equation:

$$-\frac{dL}{dT} = KL \qquad (20)$$

where,

t = time

L = volume of methane remaining to be produced after time t

k = gas production rate constant

This kinetic model is analogous to models often used to describe oxygen uptake in dilute aqueous solutions by bacteria utilizing soluble organic matter as substrate. One example of such an application of the model is reduction of biochemical oxygen demand (BOD) in the BOD bottle. Another example is the deoxygenation term of the classical Streeter-Phelps equation used to describe the oxygen deficit due to bacterial metabolism downstream from an input of organic wastes into a river. These examples illustrate that equation (20) is often considered an appropriate estimation of bacterial growth kinetics (which can be measured by uptake of essential substances or release of byproducts, such as gases) as a function of available organic substrate, when the bacterial growth system is substrate-limited.

Integration of equation (20) gives the following expressions for landfill gas production:

$$L = L_o e^{-kt}$$

$$G = L_o - L = L_o\ (1 - e^{-kt}) \qquad (21)$$

where,

L_o = total volume of methane ultimately to be produced

G = volume of methane produced prior to time t

All other terms are defined above

Convenient expressions for methane production rate are as follows:

$$\text{Methane Production Rate} = \frac{dG}{dt} = -\frac{dL}{dt} = kL = kL_o e^{-kt} \qquad (22)$$

The pertinent terms of these relationships have the following units:

$$[L] = [G] = \frac{\text{volume of methane}}{\text{mass of refuse}} \quad (\text{e.g., L/kg})$$

$$[K] = [\frac{1}{\text{Time}}] \quad (\text{e.g., 1/yr})$$

$$[\frac{dG}{dt}] = [\frac{\text{volume of methane}}{\text{mass of refuse - time}}] \quad (\text{e.g., L/kg/yr})$$

In the Scholl Canyon report's analysis of landfill gas production kinetics, the refuse mass was broken down into the sub-masses which were placed during each year of the landfill's operation. Letting the subscript i denote values for sub-mass i, an expression for composite methane gas production rate at a point in time, kL, can be written as follows:

$$kL = \sum_{i=1}^{n} r_i\ k_i\ L_{o_i}\ e^{-k_i t_i}$$

where,

n = number of sub-masses considered

r_i = fraction of total refuse mass contained in sub-mass i

t_i = time from placement of sub-mass i to point in time at which composite production rate is desired

k_i = gas production constant for submass i

All other terms are defined above

Assuming that k and L_o are the same for each sub-mass, this expression reduces to the following:

$$kL = kL_o \sum_{i=1}^{n} r_i e_i^{-k_i t_i} \tag{23}$$

From gas extraction testing, the composite methane production rate kL can be estimated (typically in the range of 1.25 to 7.49 L/kg of refuse per year for relatively young landfills). The time from placement of each refuse sub-mass, t_i, is known, and r_i is calculated as the ratio of the weight of each sub-mass to the total refuse weight. The value of L_o is estimated from the refuse composition, e.g., 62.4 to 187 L of methane per kilogram of refuse. Hence, when a composite production rate has been estimated at some point in time, the only remaining unknown in equation (23) is the rate constant, k; and equation (23) can be solved for k by trial and error. Once k has been estimated, it can be substituted into equation (21) to determine cumulative methane production or remaining methane production potential at any time, or it can be substituted into equation (22) to determine methane production rate at any time.

The estimated cumulative methane production rate is shown in Figure 6. The shaded area in the figure shows a range of methane production assuming a lower limit of methane production per unit mass of refuse of L_o = 62.4 L CH_4/kg of refuse, and an upper limit of L_o = 125 L CH_4/kg of refuse. In other words, given these assumed limits, estimated methane production would fall in the shaded area.

There is no guarantee that the Scholl Canyon approach, which is based on observed characteristics of substrate-limited bacterial growth, will accurately estimate the time dependency of landfill gas production. As already noted, many environmental factors other than overall substrate availability may be rate limiting in the landfill.

The data are not available to determine which approach is the most suitable model of landfill gas production kinetics. One may consider other reasonable gas production approaches which differ from the Scholl Canyon assumption that peak production rate is established after a negligible lagtime and is followed by a first-order (with respect to remaining methane production potential) decrease in production. For example, peak production rate may be reached quickly and the production rate may then be relatively constant for a significant period of time, followed by a gradually decreasing production rate.

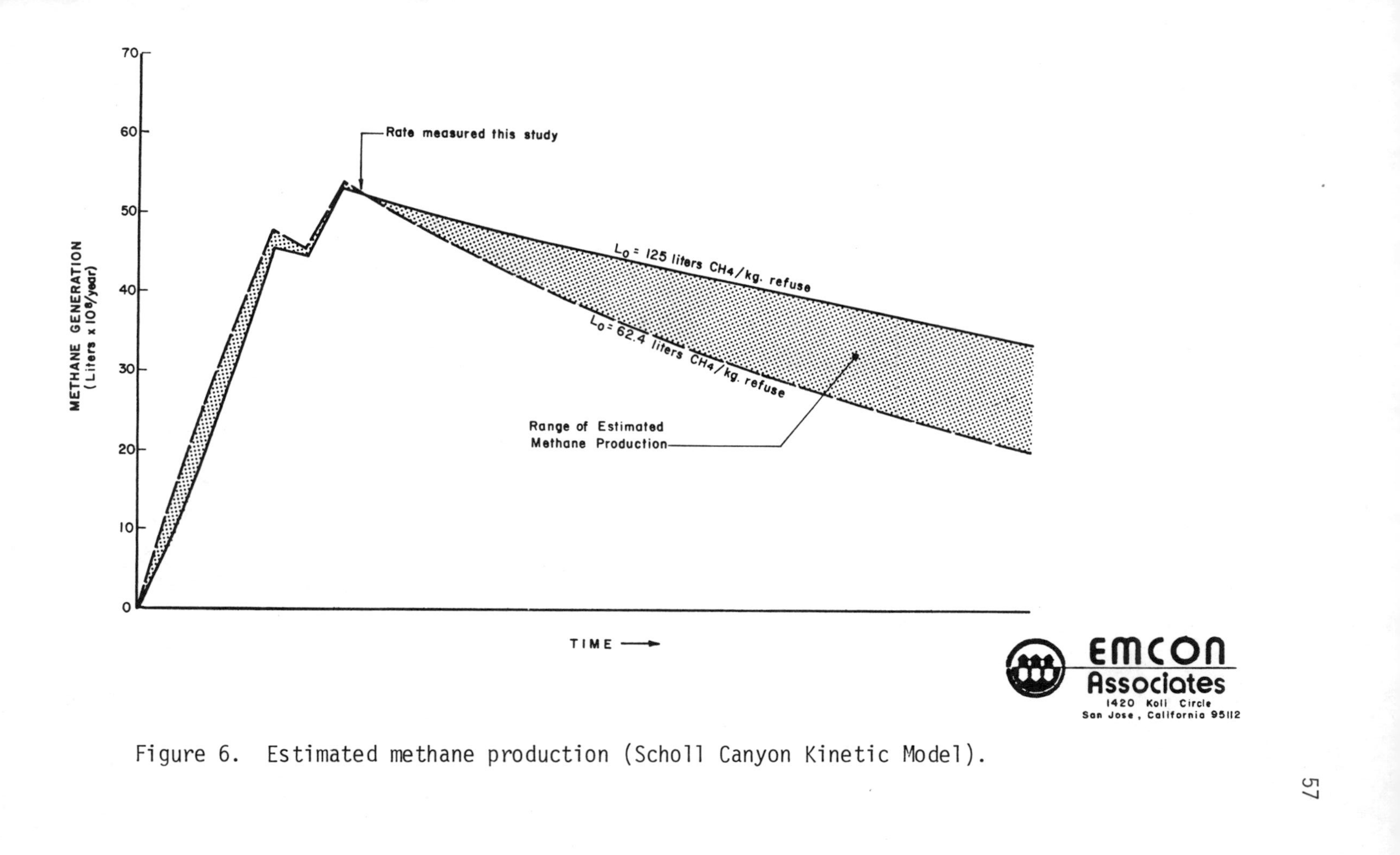

Figure 6. Estimated methane production (Scholl Canyon Kinetic Model).

Alternatively, production rate may initially increase over a significant period of time (as assumed by the Palos Verdes and Sheldon Arleta models) and gradually decrease during the later years of the landfill's gas production life (with or without a period of relatively constant production rate between the periods of increase and decrease).

GAS FLOW IN LANDFILLS

GENERAL CONSIDERATIONS

Gases flow through refuse or soils either by convection or by diffusion. Convection occurs when total gas pressure is not uniform throughout the system (i.e., when a total pressure gradient exists). Convective flow is in the direction in which total pressure decreases; gases tend to move from regions of high pressure to regions of low pressure. Diffusive flow occurs when the concentration of a gas is not uniform throughout the system (i.e., when the partial pressure gradient for a gas is not equal to zero). Diffusive flow of a gas is in the direction in which its concentration (partial pressure) decreases; gases tend to move from regions of high concentration to regions of low concentration. For a particular gas, convective and diffusive flows may be in opposing directions, resulting in an overall tendency toward cancellation. However, for most cases of practical interest related to recovery of landfill gases, diffusive and convective flows occur in the same direction.

Diffusion can occur by several mechanisms, including ordinary molecular diffusion, Knudsen diffusion, and surface migration (the latter mechanism is significant only when diffusing gases are adsorbed onto the porous medium in a mobile layer, and surface migration can probably be neglected for flow of methane in and around a landfill) [22]. Although diffusive flow is important in assessing the landfill gas hazard due to lateral migration and in evaluating passive systems for the control of such lateral migration, for most purposes its effect is negligible where an induced exhaust system is used to create total pressure gradients throughout the landfill.

In cases where gases are withdrawn from a landfill by applying a partial vacuum to wells penetrating the refuse, convective flows predominate. Pressure within the gas recovery well is reduced to some value below that normally existing within the landfill, creating a total pressure gradient which decreases in the direction of the recovery well. The net result is convective flow of gases toward the well.

MATHEMATICAL DESCRIPTION OF GAS FLOW

Darcy's law has often been used to describe laminar flow of fluids through porous media. It is most often used to describe the flow of water through porous media (e.g., the flow of ground water through a natural aquifer or the flow of water or wastewater through a sand filter during a treatment process), but it has also been applied to the flow of gases toward a production well [22]. Darcy's law applies only to laminar (as opposed to turbulent) flow; that is, inertial forces must be negligible, compared to viscous forces.

To determine whether Darcy's law is applicable to fluid flow under a given set of conditions, one can evaluate the Reynolds number--a dimensionless parameter expressing the ratio of inertial forces to viscous forces. The Reynolds number is defined by the following equation:

$$R = \frac{\rho \upsilon D}{\mu} \tag{24}$$

where,

R = Reynolds number
μ = absolute viscosity of the fluid
ρ = density of the fluid
υ = velocity of flow
D = a characteristic dimension of the system

For flow of gases through porous media, υ is taken as the apparent or bulk velocity of the fluid (as opposed to interstitial velocities, which vary in magnitude and direction throughout the system). This hypothetical velocity is obtained by dividing the flow rate by the cross-sectional area, normal to the direction of flow, across which the gases are passing. The characteristic dimension, D, is generally taken to be the mean grain diameter of the porous medium [52, 56]. Flow has been found to depart from the laminar regime as the Reynolds number exceeds some value, generally ranging from 1 to 10, depending on the distribution of grain sizes and shapes [52, 56]. Applicability of Darcy's law to gas flows resulting from a landfill gas recovery program can be evaluated on the basis of the Reynolds number and conditions likely to prevail in a landfill setting [22].

We can conservatively state that flow will be laminar if the Reynolds number is less than one. This condition, assuring laminar flow, may be expressed mathematically, as follows:

$$\text{if } \frac{\rho \upsilon}{\mu} < 1, \text{ then flow is laminar}$$

For assumed conditions of pressure, temperature, and gas composition, ρ and μ are known. If a realistic value of υ is selected, an estimate of D, the mean grain diameter of the porous medium, which assures the existence of laminar flow can be made as follows:

$$\text{if } D < \frac{\mu}{\rho \upsilon}, \text{ then flow is laminar}$$

If D exceeds the value so estimated, flow may depart from the laminar regime. The most conservative estimate of the critical value of D will result when μ is minimized and ρ and υ are maximized. The value of υ will be greatest at the refuse/recovery well interface; a realistic maximum estimate of υ at this point, for relatively high extraction rates, would be on the order of 0.3 cm/sec. Maximum ρ and minimum μ would occur at relatively low gas temperatures. Gas temperatures as low as 0°C have been observed during very cold weather. At 0°C and 1 atmosphere of pressure (actually pressure would be slightly below atmospheric because of the applied vacuum), approximate values of ρ and μ for methane and carbon dioxide are as follows:

$$\rho\ CH_4 = 0.714 \times 10^{-4}\ \text{kg/m}^3$$

$$\rho\ CO_2 = 1.95\ \text{kg/m}^3$$

$$\mu\ CH_4 = 1.04 \times 10^{-5}\ \frac{\text{N = sec}}{\text{m}^2}$$

$$\mu\ CO_2 = 1.39 \times 10^{-5}\ \frac{\text{N - sec}}{\text{m}^2} \text{ (where N is newtons)}$$

Representative values for the composite gas would fall between these extremes, depending on gas composition; using the values for CO_2 will result in a conservative estimate of the critical D, as follows:

$$\text{If } D < \frac{\mu}{\rho \upsilon} = \frac{\left[1.39 \times 10^{-5}\ \frac{\text{N - sec}}{\text{m}^2}\right] \left[\frac{\text{kg - m}}{\text{N - sec}^2}\right]}{\left[(1.95\ \text{kg/m}^3)\right] \left[(3.05 \times 10^{-3}\text{m/sec})\right]}$$

then flow is laminar

If $D < 0.233$ cm = 0.00765 ft, then flow is laminar

This result means that laminar flow is assured to exist at the refuse/recovery well interface when the mean grain size of the porous medium is less than 0.2 cm. By coincidence, this is also the grain size which distinguishes gravels from coarse sands, according to the Massachusetts Institute of Technology Grain-Size Classification System.

As a general rule, it appears that for flow rates of practical interest in landfill gas recovery, flow will be laminar and Darcy's law will apply if characteristic grain sizes of the refuse or soils are smaller than the grain sizes typical of gravel. If grain sizes of the porous medium reach those characteristic of gravel, turbulent flow may become significant in the vicinity of the recovery wells; in this case, Darcy's law would be inadequate.

It should be noted that the value calculated as the critical value of D is a very conservative estimate for the following reasons: (1) laminar flow may occur at Reynolds numbers greater than one; (2) at the modest well withdrawal rates employed in a gas recovery system, gas velocity at the refuse/well interface would probably be less than 3.05×10^{-3} m/sec; (3) gas temperatures in a landfill would normally exceed $0^{\circ}C$ (typically on the order of $38^{\circ}C$; (4) values for ρ and μ would fall between those for methane and carbon dioxide, resulting in a larger estimate for a critical value of D than that obtained by using the carbon dioxide values; and (5) this approach has estimated the critical mean grain diameter at the refuse/recovery well interface, where flow rate and velocity are maximum. As one moves radially away from the recovery well, gas flow rate and velocity across a concentric cylindrical surface surrounding the well <u>decrease</u> because gas is produced throughout the landfill, and because the area of this cylindrical surface increases. As a result, the apparent gas velocity decreases as one moves away from the well.

We can conclude that, for most cases of interest, Darcy's law is an adequate description of gas flow through refuse and through sands and silts during forced withdrawal of gases from, or adjacent to, a landfill. The principle may also be applicable to flow through fine gravels, depending on specific conditions; in addition, it is probably applicable to flow through many clays. However, Darcy's law may not be applicable to flows (1) through very coarse grained media, (2) through media containing numerous cracks and fissures, or (3) during very heavy pumping of gases, as turbulent flow is likely to be significant near the extraction wells. Also, the principle may not be applicable to extremely fine grained soils (e.g., some clays) where Knudsen diffusion may control the rate of gas transport [38].

Darcy's law for radial flow of landfill gases toward a recovery well, considering flow in other directions to be negligible, may be expressed mathematically, as follows:

$$\upsilon_r = - k \frac{dh}{dr} \tag{25}$$

where,

r = radial distance from the recovery well
υ_r = apparent gas velocity at r in a direction toward the well
k = coefficient of permeability
h = static head (piezometric head)

The derivative, dh/dr, represents the static head gradient (slope of the hydraulic grade line) at distance r. The negative sign indicates that flow is in the direction in which dh/dr decreases (i.e., toward the recovery well). The static head is expressed, as follows:

$$h = \frac{p}{\gamma} + z \tag{26}$$

where,

p = total pressure at distance r
γ = specific weight of the gas
z = elevation above some arbitrary datum

Darcy's law may now be written, as follows:

$$\upsilon_r = - k \frac{d\left(\frac{p}{\gamma} + z\right)}{dr} \tag{27}$$

Assuming that gas flow streamlines are horizontal, z is constant along a streamline, and its derivative is zero. Further assuming that the gases are incompressible over the small range of differential pressures employed in landfill gas recovery, γ can be treated as a constant and taken outside the derivative, giving the following expression for Darcy's law:

$$r = \frac{-k}{\gamma} \frac{dp}{dr} \tag{28}$$

The standard coefficient of permeability, k, depends on characteristics of both the fluid and the porous medium. This coefficient can be expressed, as follows [52, 55]:

$$k = \frac{\gamma}{\mu} k_s \tag{29}$$

where,

μ = absolute viscosity of the gas
k_s = specific or intrinsic permeability of the medium

The value of k_s is independent of fluid properties and depends on the following properties of the porous medium: porosity, range and distribution of grain sizes and shapes, orientation and packing of the grains. In terms of intrinsic permeability of the medium, Darcy's law may be expressed as follows:

$$\upsilon_r = \frac{-k_s}{\mu} \frac{dp}{dr} \tag{30}$$

Modelling gas flow through porous media requires: (1) a set of equations describing mass transport for each gas which include a term for diffusive flow and convective flow; (2) an equation describing fluid flow (Darcy's Law); and (3) an equation of state for the gases. Several attempts at modelling gas migration from sanitary landfills have been attempted, but are not summarized in detail here [36, 38].

Of the available models designed to describe gas flow through or from a sanitary landfill, two apply to gas migration outside the landfill and do not contain an explicit term for the production of gas within the landfill mass [35, 36, 38]. A third model is designed to describe gas migration within a landfill mass and incorporates an explicit term for the production of gases within the landfill volume [15]. All of the models require site-specific data such as porosity, permeability, gas composition, pressure gradients, or rate of gas production. None of the models can presently account for gas movement under a partial vacuum during pumping.

The following dimensions apply to the equations presented in this discussion:

$$p = \frac{\text{Force}}{\text{Length}^2} \text{ (e.g., 1 atmosphere} = 1.0132 \times 10^6 \frac{\text{dynes}}{\text{cm}^2})$$

$$r = \text{Length}$$

$$\gamma = \frac{\text{Force}}{\text{Length}^3}$$

$$\mu = \frac{\text{Force - Time}}{\text{Length}^2} \text{ (e.g., 1 poise} = 1 \frac{\text{dyne sec}}{\text{cm}^2})$$

$$k = \upsilon_r = \frac{\text{Length}}{\text{Time}}$$

$$k_s = \text{Length}^2 \text{ (e.g., 1 darcy} = 9.87 \times 10^{-9} \text{cm}^2)$$

FIELD TESTING OF GAS RECOVERY FROM LANDFILLS

GENERAL CONSIDERATIONS

Sampling and Monitoring Devices

Bar-Hole Probe

The simplest method of monitoring gas composition in soils and refuse is by means of a bar-hole probe. The probe is simply a rigid, hollow tube which is attached to the inlet of a gas detection device by means of flexible tubing. The probes are typically 60 to 90 cm long and are used in conjunction with a bar-hole driver. The latter, consisting of a small diameter solid metal rod with a slide-hammer attached, is used to make an opening in the soil to accommodate the bar-hole probe. The probe is inserted into the bar hole, and the hole is sealed around the probe at the surface, typically with rubber stopper, cloth, or native soil. A gas sample is then drawn from the probe and through the gas detection instrument, giving an indication of the gas composition in the bar hole.

Detection of methane by means of a bar-hole probe may give a positive indication of the presence of combustible gas, but a failure to detect methane does not necessarily indicate the absence of combustible gas. Most frequently, the gas sampling is attempted immediately following the creation of the probe hole. The accumulation of migrating gas to a representative level may require minutes, hours, or days. In numerous investigations, shallow bar-hole probe surveys have failed to detect significant methane concentrations, and subsequent monitoring at deeper, permanent probes has shown the gas to be present at relatively high levels. Another limitation of the bar-hole probe survey is the fact that results are not repeatable because a new hole must be made each time a survey is conducted. For these reasons, installation of permanent gas monitoring probes with periodic monitoring over a period of time is preferable to bar-hole probe surveys when investigating the presence and characteristics of landfill gas in soils or refuse.

Permanent Gas Monitoring Probes

Installation of a permanent gas monitoring probe requires drilling a hole in the soil or refuse to a depth at which monitoring is desired. The probe casing, perforated at the tip, is installed in the drilled hole, and the hole is backfilled with permeable material (sand or pea gravel) to a distance above the perforations. The remainder of the hole is backfilled with fine-grained soil to act as a seal against the instrusion of air from the surface. Impermeable seals are sometimes warranted to keep soil above from entering the permeable material. A gas sample can then be withdrawn from the probe at the surface, or the probe pressure can be determined. Probe measurements must be taken at the probe tip if they are to be representative of gas composition and internal landfill pressure at specified depth.

Figure 7 illustrates a typical probe installation and Figure 8 presents details of a gas probe tip.

A thin-walled PVC (polyvinyl chloride) pipe 1.3 cm in diameter is commonly used for the probe casing. To obtain a sample of gas representative of conditions at the probe tip, a gas volume equivalent to the entire volume of the probe between the probe tip and top of the pipe must be withdrawn. Since methane is lighter than air and carbon dioxide is heavier, some stratification of the gas is frequently observed in the probe. A probe having an internal, flexible tubing of small diameter is recommended since it requires less gas volume extraction to obtain a representative sample at the tip. This is a very important consideration when gas production is slow, and/or when there is a negative pressure at the probe tip (e.g., due to diurnal pressure fluctuation or exhaust testing).

Gas Extraction Wells

The gas extraction wells used for landfill testing are similar to the wells used for full-time operation of permanent recovery and/or control systems. Through careful planning, test wells used to determine the feasibility of methane recovery can later function as part of the permanent control well installation. Well construction will be discussed in a later section in some detail.

Under static conditions, the extraction wells are used essentially as monitoring probes. During short-term extraction

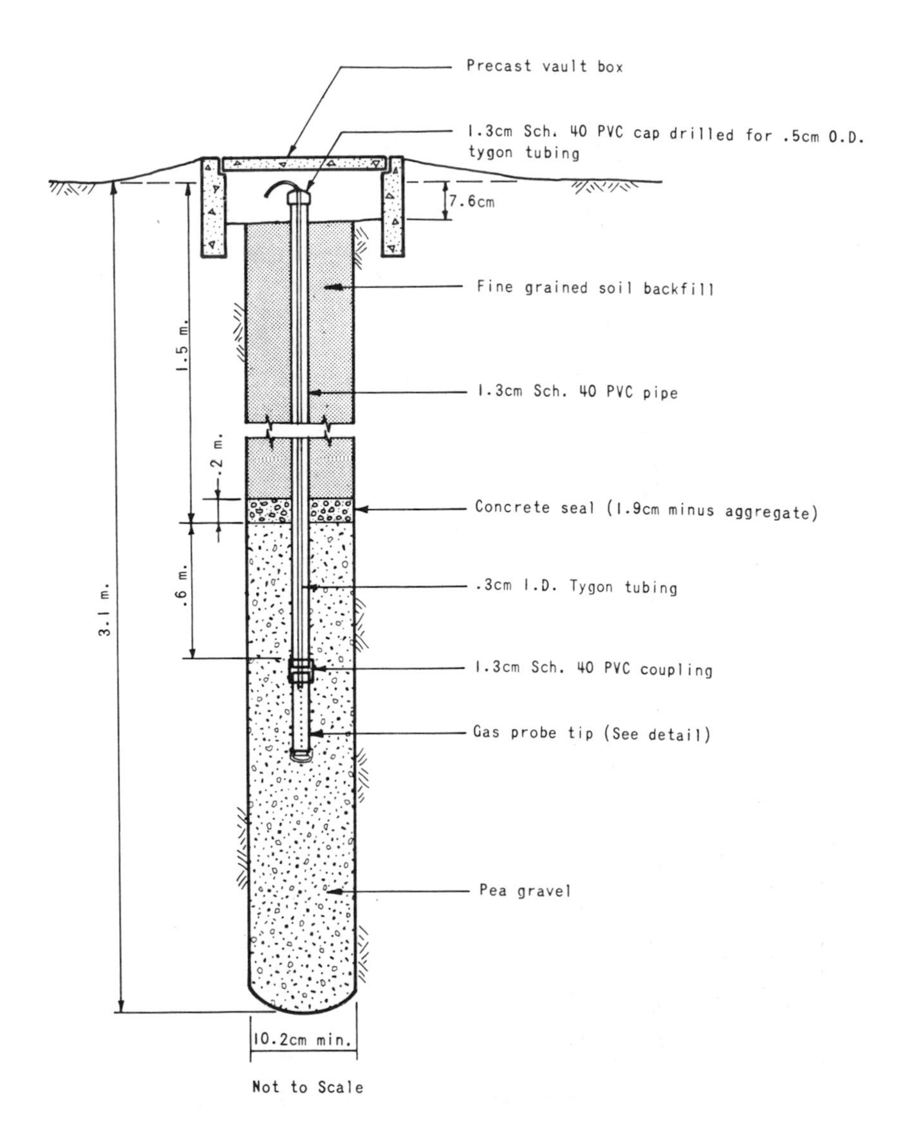

Figure 7. Typical probe installation.

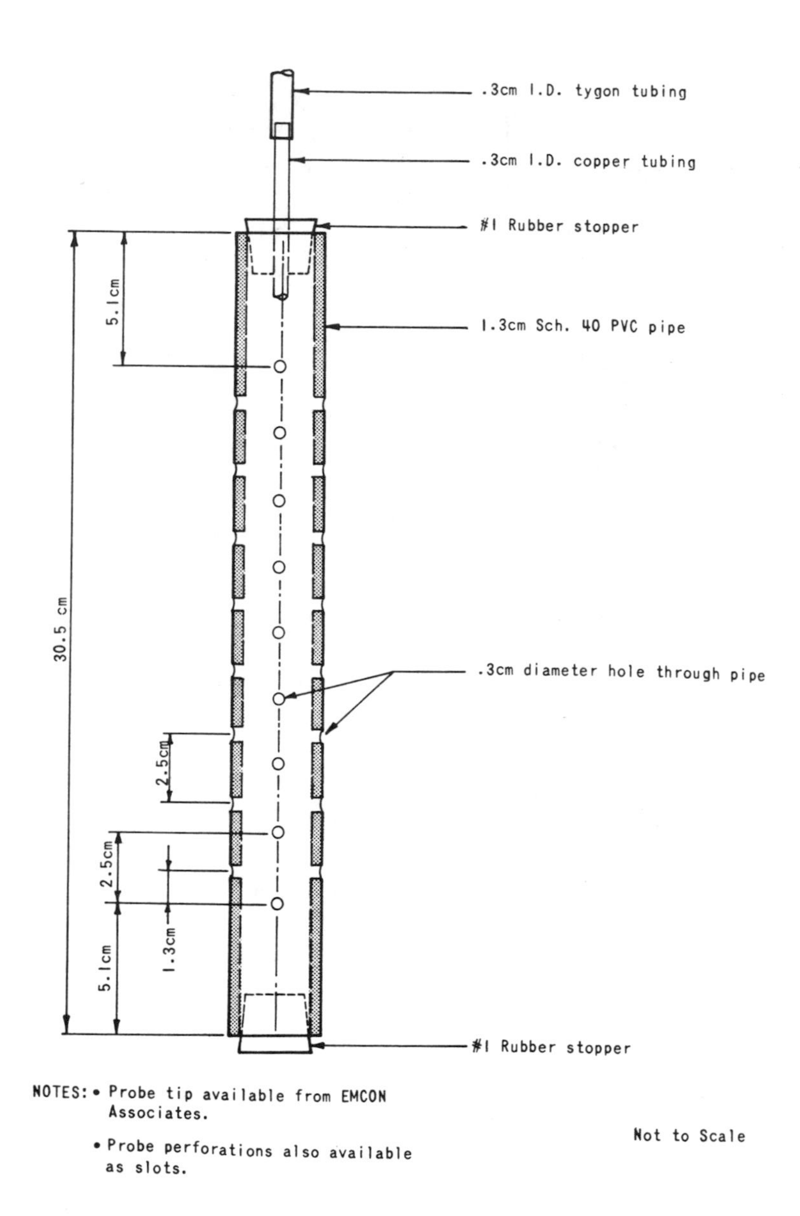

Figure 8. Detail--gas probe tip.

testing, gas is extracted from wells singularly or in combination by means of portable blowers. During long-term extraction testing, gas is frequently extracted simultaneously from several wells which are linked by means of gas collection header piping. A motor/blower unit applies a suction to the collection header, gathering the gas to one point where it is exhausted. A burner is often needed to flare the exhaust gas in order to control malodours. Details regarding the gas collection header and blower/burner facility are presented in a later section.

Gas Characteristics Determined by Field Testing

Gas Composition

Gas composition is one of the determinations that is of principal interest in any testing program. Field instrumentation is readily available for routine determination of combustible gas, carbon dioxide, and oxygen concentrations. Verification of the accuracy of field determinations and a more detailed breakdown of gas composition can be made by collecting a gas sample and submitting it for laboratory analysis. Typically, methane, carbon dioxide, oxygen, and nitrogen concentrations are determined with a laboratory gas chromatograph; however, there is an increasing use of portable field chromatograph equipment. Concentrations of hydrocarbons other than methane and those of trace gases that may have an effect on usability (e.g., corrosive compounds) can be determined by mass spectroscopy.

The significance of methane concentrations is apparent in that the energy content of landfill gas is directly proportional to its methane concentration (assuming other combustible gas concentrations to be negligible, which is generally the case). The presence of oxygen or nitrogen is usually an indication of air intrusion; however, small amounts of nitrogen may be generated within the landfill, especially when the fill is young. Moisture content and the presence of corrosive compounds is important when evaluating the extent of gas processing required in a gas recovery project. Table 8 presents some typical landfill gas compositions, as observed at various landfills.

Relative Pressure

The production and accumulation of gases as end products of the anaerobic biological activity within the landfill tends to raise the gas pressure in the fill voids. Under static conditions, internal landfill pressures higher than atmospheric are normal. The resulting gradient in total pressure from fill to the surrounding soils and the atmosphere is one of the driving forces (along with diffusion) creating a net outflow of landfill gas away from the landfill. Static pressure measurements have generally been found to exhibit a diurnal fluctuation, reaching the highest and most stable internal pressures during the afternoon hours. This diurnal fluctuation is due to changes in atmospheric pressure as well as to changes within the landfill. The magnitude of the internal, relative pressure depends, in part, on the nature of the surrounding soils and the landfill cover. Less permeable soils and cover can be expected to result in higher internal pressures, all other factors being equal. Static measurements at numerous landfills have revealed internal, relative pressures generally on the order of a few millimetres of mercury or less, although values as high as 225 mm of mercury have been observed in isolated locations at landfills where the average internal pressure is less than 2 mm Hg.

During gas extraction from a landfill, the pressure distribution around the withdrawal wells defines the well's influence on the refuse mass. The pressure distribution discovered during extraction testing can be used to determine the layout of a full-recovery well system.

Pressure readings (along with gas composition determinations) made at shallow probes during gas extraction are useful in evaluating the extent of air intrusion through the cover. Pressure below the cover should be kept slightly higher than atmospheric pressure to prevent convective flow of atmospheric oxygen and nitrogen into the fill.

Pressure measurements made within the casing of the extraction wells during gas test withdrawals are helpful in correlating the well flow rates with the applied suction. This information can be helpful in sizing the gas collection header piping and the motor/blower unit.

Temperature

The biological activity within the landfill results in heat, part of which is carried away by the gases generated. Therefore, elevated gas temperatures are a sign that biological activity is taking place within the landfill. Landfill gas temperatures typically range from 30°C to 40°C. Shallow landfills may exhibit near-freezing temperatures during cold winters. Landfills of at least 15 m in thickness are relatively unaffected by ambient air temperatures, and have been observed with temperatures as high as 70°C. Such high temperatures may exist in "pockets" or microenvironments within a landfill whose average temperature may be 40°C.

Flow Rate

Because internal landfill pressures are usually higher than ambient, atmospheric pressure, convective gas flow out of an uncapped monitoring probe or extraction wells is typically observed under static conditions. The magnitude of the flow rate can be compared to past measurements made at other landfills using similar probes or wells, to estimate the relative gas production potentials. These gas flow rates are determined by a combination of many factors, including hole diameter, interception length and depth, differential pressures (atmospheric versus landfill) and all other parameters affecting landfill gas generation. Diurnal pressure variations can explain why a probe will support combustion of a flame in the afternoon and evening, but not necessarily in the morning when internal pressure may be less than atmospheric.

During extraction testing, gas withdrawal flow rates must be regularly determined, to be used (along with gas composition and pressure distribution data) in estimating production and recovery potential.

Available Field Equipment and Instrumentation

Gas Composition

Field determination of combustible gas and oxygen concentration is normally made with portable gas detection instruments. These instruments can be obtained for making measurements in numerous ranges of methane concentration, ranging from a few parts per million to 100%. The more sensitive units are most useful for trace gas studies such as leak detection when small concentrations are present, and for trend studies of small concentrations. The gas analyzer units are more helpful for gas hazard and recovery studies; these units may have one meter for the 0 to 100% lower explosive range (LEL) and 0 to 100% combustible gas. The instruments often make use of a Wheatstone Bridge circuit which senses changes in an electrical coil's conductivity with changes in temperature caused by flowing the gas over the coil and/or combusting it. The units may be equipped with an electrical pump, or the sample may be drawn with a hand-aspirated bulb. The electrical pump system is far more convenient for field use.

Other equipment of value in the field includes: (1) combustible gas analyzers with oxygen meter; (2) portable, relatively low-cost gas chromatographs; (3) infra-red detectors for methane-carbon dioxide analysis; and (4) orsat-type carbon dioxide and oxygen indicators adapted from the field of chemical engineering. The orsat-type units are relatively inexpensive and easy to operate, but their usefulness is questionable because the data they provide usually duplicate those obtained by other methods.

Sample Containers

Gas sample containers are readily available in glass, stainless steel, and plastic material. They typically are in the 250 mL size with stopcock valves at either end. Selection of material is based on consideration of use, cost, breakage, transportation, and life. The plastic sample containers are acceptable and should be the preferred selection in most instances.

Laboratory analyses can be performed on samples of gas obtained in the field. Gas sampling bottles are used for collection and storage of the gas. The bottles are simply pressure bottles equipped with an opening and stopcock on each end. Hand-operated or battery-powered pumps may be used to vacate the vessel before sampling or to pressurize the gas, forcing it through and into the vessel. Alternately, fluid displacement techniques can be employed to obtain a representative gas sample.

Pressure

Gas probe and well pressures must be obtained in the field. An airtight connection must be made between the probe, or well, and the pressure indicator. The simplest pressure instrument in common usage is the inclined manometer. Pressure readings accurate to ±0.002 mm of Hg can be obtained using a manometer equipped with a micrometer. An alternative to the manometer is the Dwyer, Magnehelic differential pressure gauge. The range of these gauges is from 0-0.47 mm Hg to 0-281 mm Hg, and they are accurate to ±2% of full scale. The Magnehelic gauges have been found to be quite convenient and reliable in field applications. They should be tripod-mounted for ease of use in the field, as they must be calibrated by leveling prior to each measurement.

Temperature

Mercury-filled glass thermometers have been used to obtain landfill gas temperatures; under static conditions, these devices are of limited utility in obtaining temperatures at wells. The thermometer must be used near the surface and results may be affected by ambient temperatures. During gas extraction, the thermometers give acceptable measurements of temperature of the gas flowing through the well head and in the collection header. For measuring probe-tip and static-well tip temperatures, thermocouples or thermistors are preferred; these can be lowered into or permanently stationed at the tip, relaying temperature data via electrical cable.

Flow Rate

Flow rates may be estimated by using the continuity equation: flow rate equals area normal to flow times average gas velocity. Numerous devices are available for velocity measurement, including those utilizing hot-wire anemometers or rotating anemometers. But the most common means of determining velocity is with the pitot tube, used in conjunction with a manometer or Magnehelic gauge. By making only one measurement in the center of the conduit with a pitot tube, one may achieve accuracies of ±15%.

DESIGN OF FIELD TESTING STUDY

Background Information

Operating History of the Landfill

Planning the field test is facilitated by general historical information related to the production and recovery potential of the landfill under study. For example, categorizing the types of refuse accepted for landfilling gives a clue as to refuse composition, permitting comparative estimation of total gas yield per unit mass of refuse. This refuse categorization may also help assess the potential impact of toxic substances. Other operational information may help develop a general idea of whether a landfill should be a good methane producer. For example, a landfill at which moisture was added to aid in the compaction of refuse would be expected to produce methane at a greater rate (at least during its early life) than would a relatively dry landfill. Data concerning placement tonnage and chronology help one to evaluate a landfill's gas production potential, while the fill's age and location distribution facilitate the layout of the test wells and probes. Conditions of the surrounding soils and the cover material (especially information related to permeability) help to determine desirable well setbacks from the landfill perimeter, inter-well spacings, and test extraction rates.

Refuse Analyses

Refuse composition is an important factor limiting both the total yield and the rate of production that can be expected from a landfill. During refuse placement, the incoming refuse can be randomly sampled and hand sorted into categories, such as paper, yard waste, food waste, wood, textiles, leather, rubber, plastics, and inorganic materials. The percentage distribution by weight of these categories of refuse, compared to similar data from other landfills, gives an indication of a landfill's gas production potential.

In addition to the above compositional analysis, certain gross characteristics more convenient to obtain can be used to indicate a landfill's methane generation potential. Refuse moisture and volatile solids determinations are useful indicators, higher values being associated with greater gas production potential. Total organic carbon content is another relative measure of production potential. These gross characteristics can be determined for raw refuse being placed, as well as for refuse samples withdrawn from the landfill during drilling operations. A drilling log of refuse appearance can also be of value, especially in evaluating differences between two locations within the same landfill.

Objectives of Field Testing

Static Testing

A general idea of a landfill's potential as a methane producer can be formed by evaluating a site's physical characteristics and operational history, and by comparing gas composition, pressure, temperature, and flow data under static conditions with similar data obtained from other landfills. Diurnal and seasonal fluctuations in a landfill's gas production patterns can be discovered by static monitoring. The static test results are reviewed to determine whether extraction testing may be worthwhile. Most importantly, data gathered under static conditions serve as a baseline against which data obtained during extraction testing can be evaluated; in particular, establishing a static baseline is an essential prerequisite to determining the effect of gas extraction on internal landfill pressures.

Short-Term Extraction Testing

Short-term extraction testing is a means of obtaining a first look at pressure and gas composition distributions surrounding an extraction well, as they relate to the extraction rate employed. The results of short-term testing give an indication of the extent of air intrusion, and the influence of a selected withdrawal rate on the refuse mass. In addition, the applied vacuum required to achieve a given flow rate permits more rational selection of desirable well spacings, blower capacity, and well-flow rates. Preliminary estimates of gas production and recovery rates can be made on the basis of short-term testing.

A landfill's volume can be expected to include about 50% void space. Because landfill gas can be stored in the void space, the results of short-term testing may not reflect the current gas production rate. Short-term testing may be more analogous to the withdrawal of gas from a reservoir than to conditions required for an ongoing landfill gas recovery project. In evaluating the results of short-term testing, it is difficult to separate the portion of gas extracted that was stored in the voids at the beginning of testing from that portion which is "newly produced" (i.e., produced during extraction testing).

Short-term extraction tests usually run for periods of several hours to several days for each gas extraction rate. Typically, 2 to 4 extraction rates are used for each well, and 2 to 4 wells are tested per landfill. The main limitation of a short-term test is that only a portion of the gas being extracted is newly produced. Weeks or months of gas extraction are necessary to develop data to verify a stable, sustainable extraction rate.

Long-Term Extraction Testing

The primary objective of long-term extraction testing is to refine and verify gas production rate estimates made from the results of short-term testing, and especially to permit better estimation of the "sustainable" production or recovery rates. The long-term testing is intended to simulate conditions expected from a full recovery project. The objectives of long-term testing should be to estimate the "sustainable" gas recovery rate (or at least to establish a lower limit on the same), and to refine preliminary design criteria obtained from short-term testing. With careful planning, many of the components of the long-term test gas extraction system can be used as part of a full recovery system, if full recovery proves feasible.

Layout of Extraction Wells and Monitoring Probe for Field Testing

Area Coverage

Initial extraction testing may be from only one well; however, it is generally preferable to extract gas from various locations within the fill, especially during long-term testing. Considerable variation in conditions affecting gas production is often observed within a given landfill. For extraction testing to be representative of a field's overall gas recovery potential, wells should be located to withdraw gas from fill regions spanning the range of site conditions reflecting the more important factors affecting gas production (a discussion of these important factors was presented earlier in this report).

Several criteria are particularly valuable in locating extraction wells for field testing. If well clusters are utilized they should be widely spaced over the landfill surface area. In addition, the well locations should be selected to withdraw gas from areas covering the extremes of conditions such as refuse composition and age, and landfill characteristics and operating conditions (e.g., moisture addition). Test wells should also be located with forethought as to their eventual inclusion as part of a full recovery system.

If the short-term test results are favorable, the long-term test well layout should consider the advantage of combined pumping from the short-term wells and the addition of new wells. The cost of a combined extraction system may be lessened if wells are closely spaced and a common header-blower-burner system is utilized.

Gas monitoring probes are most commonly located along imaginary lines radiating outward from the recovery wells. Typically, the nearest probe might be 3 to 9 m from the extraction wells, and successive probes might be spaced out at probe-to-well distances approximately double the prior distance (e.g., 3, 6, 12, 25, 50, 100 m). Additional probes might be located near the landfill perimeter to evaluate air intrusion across the soil-refuse interface. Probes may also be located at other points of interest, such as at the mid-distance between adjacent wells. An example of a typical well cluster is presented on Figure 9.

Depths

Extraction wells can be located at various depths, depending on the refuse stratum from which gas extraction is desired.

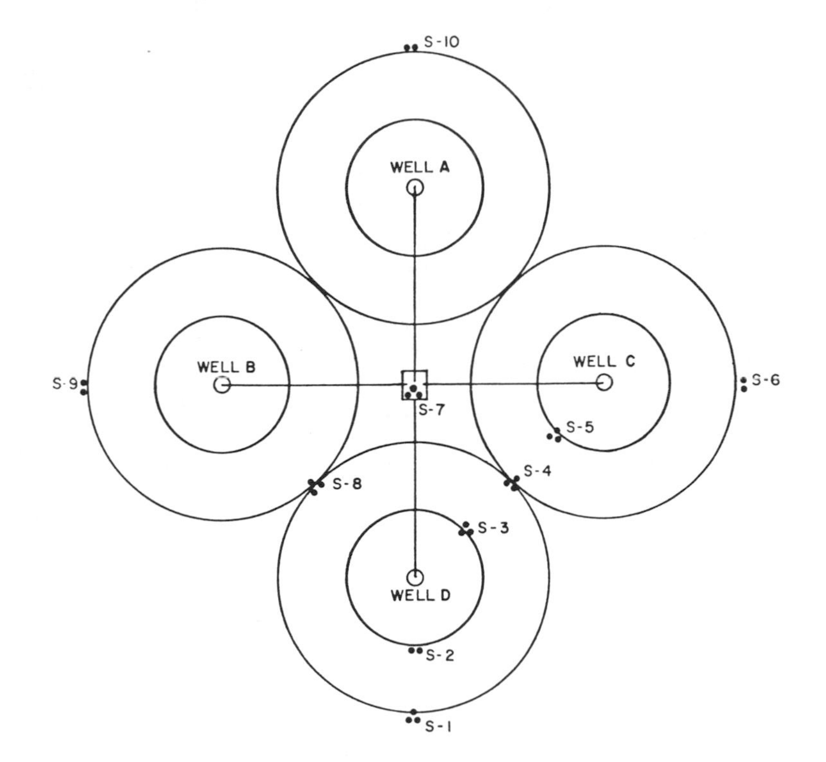

PROBE SCHEDULE

STATION	SHALLOW PROBE @ 0.2 D.	MIDDLE PROBE @0.4 D	DEEP PROBE @0.8D
S-1	X	X	X
S-2	X	X	
S-3	X	X	X
S-4	X	X	X
S-5	X	X	X
S-6	X	Optional	
S-7	X	X	X
S-8	X	X	X
S-9	X	Optional	
S-10	X	Optional	

SCALE: 1" = 200'
(Variable)

Figure 9. Well/probe cluster configuration.

More than one well casing may be placed at different depths in the same borehole, with the gas extraction intervals sealed off from one another. These multiple-depth wells permit one to evaluate (1) which refuse lifts are the biggest contributors to gas production, (2) the lifts' relative permeabilities, and (3) the relationship between apparent horizontal and vertical permeability. Typical depth of a single-casing extraction well might be 60 to 90% of the refuse thickness, with the lower 30 to 80% of the well casing perforated. The percentages chosen as design criteria would depend primarily on refuse thickness and cover conditions. For example, a shallow landfill or one with a shallow or permeable cover would call for a smaller percentage length of casing perforated than would a deeper landfill or one with a tight cover.

Probe depths generally should be varied in order to determine the depth distribution as well as the radial distribution of internal pressures and gas composition surrounding gas extraction wells. Shallow probes provide data on the extent of air intrusion through the landfill cover, whereas all probes (shallow and deep) are used to estimate the extent of an extraction well's radial and vertical influence on the refuse mass. As with extraction wells, multiple-depth probes may be placed in the same borehole, with the perforated intervals effectively sealed off from one another. Normally, a length of only 1 m or less is perforated at the probe tip. Example probe depths might be at a depth of 2 m, 6 m, and at depths of 20, 40, 60, and 80% of the average refuse thickness. Cost considerations normally preclude placing monitoring probes at each depth of interest for every probe location. Locations for multiple-depth probes might be at several radial distances from each well.

Test Program

Duration of Testing

As a general rule, greater reliability in estimates made from test results can be achieved for tests of longer duration than for those of shorter duration. Of course, cost is usually the factor which limits test duration, and there is a point beyond which the cost of extending a testing program outweighs the information that remains to be gained.

Although the duration of static testing is rather arbitrary, monitoring should continue for a period adequate to reveal diurnal trends and to establish a recognizable data baseline. Investigation of seasonal variations would require monitoring over a period of at least one year; however, the cost of monitoring for such an extended period is often prohibitive. Typically, the static test results would be monitored for a period of 2 to 5 days just prior to extraction testing. Barometric pressure data should be collected concurrently and utilized to evaluate the atmospheric impact on the measured pressures.

Short-term extraction testing typically proceeds for a period ranging from several hours to several days per extraction rate per well. Landfill pressures have been found to respond relatively quickly to the inception of extraction testing or to changes in extraction rate (stable pressures will generally have been established within a matter of a few hours or less). Changes in gas composition and long-term trends of pressure distribution require extraction of longer duration.

For a test to be categorized as "long term", it must proceed for a period of time at least long enough to extract a volume of gas equivalent to the landfill's total void volume. Estimation of the "sustainable" gas recovery rate or the establishment of a lower limit on this rate requires gas extraction for some period of time exceeding the minimum duration required to exhaust all stored gas. The test duration required to permit reasonable estimation of the sustainable recovery rate depends on the landfill void volume, the methane extraction rate, the actual potential recovery rate, and the fraction of the total refuse mass from which gas is being withdrawn, as will be noted in a later section of this report. Accurate estimation of the sustainable recovery rate requires extraction at a rate and for a period of time sufficient to (1) withdraw all stored gas, and (2) establish the maximum rate at which extraction can proceed without causing significant air intrusion or deterioration of gas quality. Tests of shorter duration and/or at lower extraction rates can be used to establish a lower limit on the recovery rate that could be sustained; however, these tests are not conducive to estimation of the actual, potential recovery rate.

Frequency and Timing of Monitoring

Determination of diurnal fluctuations in relative pressure internal to the landfill requires monitoring throughout the course of the day--for example, at 1-hour intervals. It has been observed that relative pressures generally exhibit their highest and most stable values during the afternoon hours. Baseline pressure and gas composition data should be collected under static conditions on a regular basis for some period of time prior to extraction testing. For example, static data might be obtained prior to extraction testing, daily or every other day. To obtain representative data and to make the baseline data as repeatable as possible, static readings should be made during the highest and most stable pressure time period, usually in the afternoon and at approximately the same time each day.

The frequency of monitoring during short-term testing depends largely on the test duration. For example, if an extraction test is to proceed only three hours, it may be possible to make only one round of readings, depending on the number of probes monitored. If an extraction test is to proceed for a period of a day or more, several rounds of readings might be made each day. At least one round of readings should be made each time the extraction rate is changed. As a general rule, a waiting period of at least one hour might be used after the adjustment of extraction rate before readings are taken, to permit landfill pressures to stabilize. A subsequent set of readings must repeat the previous pressure readings before the pressures are considered stable. The best method of comparing extraction data with baseline, static data is achieved by taking extraction readings at approximately the same time of day for which the most stable baseline data were developed.

Many of the comments regarding the frequency and timing of monitoring during short-term extraction testing also apply to long-term testing. Readings should be made most frequently at the beginning of testing and after any abrupt changes in conditions, such as an increase or decrease in extraction rate. During the early stages of long-term testing, a full round of readings is typically made each day, during the afternoon. As gas extraction rates and pressure and gas composition data become relatively constant, the frequency of monitoring and/or the number of probes monitored each day might be decreased in order to minimize costs. More frequent and complete monitoring should again be conducted as a total volume of extracted gas equivalent to the landfill void volume is approached.

Extraction Rates Employed

Of course, one of the objectives of extraction testing is determination of the rate at which methane gas can be extracted from the landfill. But in ordering equipment (e.g., motor/blowers) and in setting the initial operating conditions of the testing program, preliminary estimates of well-flow rates are required. A simple approach utilizing the concept of "radius of influence" can be used to obtain a reasonable estimate of well-flow rates to be expected.

The radius of influence is assumed to define a distance from the extraction well within which all gas is drawn to the extraction well; it is assumed that no gas is drawn to the extraction well from a distance greater than the well's radius of influence. Within this context, the radius of influence is a fictitious concept in that not all the gas within the radius of influence can be captured (i.e., some gas will escape to the atmosphere) and some gas will be drawn to the well from a distance greater than the radius of influence. However, the errors inherent in these two simplifications tend to cancel one another, and reasonable flow rate estimates are usually achieved. The concept of radius of influence is a convenient and useful one in establishing inter-well spacings and in estimating potential well-flow rates.

Using the radius of influence approach, an individual well-flow rate (Q_w) can be estimated as follows:

$$Q_w = \frac{K \pi R^2 tDr}{C} \tag{31}$$

where,

Q_w = well-flow rate, L/sec

K = a compilation of conversion factors, 1.157×10^{-8} (L/day)/(mL/sec)

R = "radius of influence", m

t = refuse thickness, m

D = in-place refuse density, kg/m^3

r = methane production rate, mL/kg/day

C = fractional methane concentration

Some of the values needed to estimate Q_w will be available for a given landfill; others must be estimated or assumed. For example, assuming the following typical values:

D = 650 kg/m^3

r = 7 mL/kg/day

C = 0.5

equation (31) reduces to the following:

$$Q_w = K'R^2 t \tag{32}$$

where,

$K' = 3.31 \times 10^{-4}$ $L/sec/m^3$ refuse.

Refuse thickness is usually known. Q_w and R are interrelated, requiring the selection of one or the other.

If several wells are being used for extraction, the desired radius of influence might be selected as approximately one-half of the inter-well distance. For example, if refuse thickness is 30 m, the extraction rate required to establish a radius of influence of 70 m would be estimated from equation (32) at 49 L/sec; and reasonable test extraction rates for a group of extraction wells spaced about 140 m apart might range from 20 to 125 L/sec, depending on the validity of the assumptions. For extraction from a single well, extraction rates considerably greater than the above estimate could be employed, as the extent of the radius of influence would not be limited by the proximity of adjacent wells. The range of reasonable estimates noted above for the example case was calculated simply by assuming that the rate would vary between 40 and 250% of the calculated estimate. Since extraction testing is intended to place an estimate on the actual production or recovery rate, motor/blower equipment providing a range of possible extraction rates should be utilized.

The above approach might be used to estimate the required capacity of motor/blower equipment to be used for short-term testing. Normally, variations in extraction rate are used for both short-term and long-term testing. Initial high extraction rates may inadvertently overstress the landfill, causing a significant influx of air and diminishing the landfill's capacity as a methane generator. Therefore it is preferable to begin the testing with lower extraction rates and to increase the rate as testing progresses. Changes in observed pressure and gas composition should be noted as extraction rates are varied; these changes give an indication of the effect on the landfill of the extraction rate variation, permitting rational management of the testing program on the basis of the monitoring data. Results of short-term extraction testing give a starting point for extraction rates to be employed during long-term testing. However, differences in the landfill response may be observed as the gas stored within the landfill's voids is diminished.

Potential Field Testing Problems

Numerous practical problems may arise in the course of landfill gas extraction testing. For example, in installing the extraction wells and monitoring probes, difficult drilling conditions are often encountered and progress may be slow. Inclement and cold weather can hamper mobility on-site.

Equipment and instrumentation failures are potential problems that should be addressed during the planning stages of a testing program, and an ongoing maintenance and calibration routine should be established. Equipment and instrumentation are often subjected to harsh conditions, and landfill gas is always wet and sometimes corrosive. To insure that accurate and meaningful data are obtained, proper maintenance and calibration are imperative. Equally important is the training and supervision of the personnel who are to operate and maintain the equipment and who use the instrumentation in the field.

Since unexpected difficulties are almost certain to arise during the course of extraction testing, it is essential that personnel are available who can assess the situation and recommend corrective actions. Special safety precautions should be implemented for those working around the drilling and testing equipment, as the gases are flammable and explosive and deep open holes may be temporarily present.

As test extraction rates are increased, it can be anticipated that a point will be reached at which air "breakthrough" will occur (i.e., atmospheric air being drawn into and through the landfill). The breakthrough usually can be detected from changes in composition of the gas being extracted. The nitrogen concentration will increase since nitrogen can be drawn through a landfill essentially unaffected; the oxygen concentration may increase but usually does not since oxygen would be utilized by aerobic or facultative organisms near the point of air intrusion. In addition, an increase in the ratio of carbon dioxide to methane concentration may be observed since the organisms utilizing the oxygen drawn into the fill will produce carbon dioxide but no methane. Since probe readings can also indicate air breakthrough, the above comments regarding gas composition changes apply to evaluation of the probe readings as well. In addition, relative pressures at the shallow probes might be observed to drop below zero if air is being drawn through the landfill cover. If air breakthrough is observed, the extraction rate should be immediately diminished. Oxygen is toxic to the strictly anaerobic methanogenic bacteria, and some methane generating capacity of the landfill will be lost, at least for a short period of time, if air continues to be pulled through the landfill.

Problems arising from the moisture contained in landfill gas are also common. The accumulation of water at low points in gas collection header piping results in partial or complete blockage of gas flow. In addition, free water and particulate matter contained in the gas may interfere with velocity measurement by pitot tube. Periodic tapping of the tube on the interior pipe wall will help dislodge the blockage. To minimize problems associated with gas moisture, the collection header must be properly sloped and condensate traps must be constructed at the low points in the system (see section entitled "Gas collection Header" for a discussion of condensate traps).

Another problem may arise from the malodours associated with trace amounts of organic gases that are usually present. Depending on the use of adjacent property, these odours may not be a nuisance, at least for short-term testing. However, when the malodours are a problem, igniting the gas in a flare unit is the simplest method of odour control.

Field methane instrumentation should be calibrated before each day's reading by analyzing a gas of known composition obtained from a local gas supplier.

INTERPRETATION OF FIELD TESTING DATA

Refuse Analyses

Refuse composition information can be used to estimate theoretical maximum gas yield by the methods outlined earlier in this report. The estimation method that can be employed depends on the degree of detail in the available composition information.

Gross characteristics such as refuse moisture, volatile solids and total organic carbon content are useful in estimating maximum yield, and they can also help indicate whether a landfill's gas production rate should be greater or less than rates observed at other landfills. For example, all other factors being equal, a landfill containing refuse with a relatively high moisture content can be expected to produce gas at a greater rate than can a drier landfill. For a "typical" landfill, the moisture content of refuse at the time of placement is usually about 25%.

Static Testing

Static testing not only provides baseline data to be used in conjunction with extraction testing, but enables an assessment of a landfill's methane production potential. There is no rigorous mathematical approach for estimating production potential from static data; the most appropriate means of evaluating static data would be to compare the observations with static data obtained at other landfills where extraction testing has also been conducted and to develop a qualitative idea of production potential on the basis of this comparison. A landfill generating methane at a substantial rate would be expected to exhibit consistently positive internal relative pressures, relatively high gas temperatures and methane concentrations of 50% or more.

Diurnal fluctuations in static data, particularly in relative pressure, may be an indication of the consistency of gas production. In evaluating relative pressure variations, the pressure measurements should be compared to fluctuations in ambient barometric pressure. Since atmospheric pressure is the datum for relative pressure measurements, landfill pressure fluctuations may simply reflect atmospheric changes.

Static testing can also be used to evaluate seasonal variations in production potential, which may limit the feasibility of a gas recovery and use program. Seasonal variations are of particular interest for shallow landfills located where winters are cold since the biological process by which the methane is produced is temperature dependent, slowing in rate as temperatures drop.

Extraction Testing

Gas extraction provides the most useful type of data for estimation of a landfill's recovery potential. Many of the same methods of data interpretation can be used for both short-term and long-term testing. Preliminary conclusions based on short-term testing results can be refined or verified by long-term testing. Long-term testing results in greater confidence of "sustainable" methane recovery rate estimates since gas stored within the landfill voids is depleted.

Influence

During extraction testing, gas is withdrawn from a test well at known rates, and the pressure responses at pertinent monitoring probes (along with other relevant parameters) are monitored. To correct for probe-to-probe pressure variations under static conditions, fractional pressure drop relative to static pressure is the parameter commonly employed. This fractional pressure drop is referred to as influence (I), and is defined as follows:

$$I = \frac{\text{(static pressure)} - \text{(pressure during pumping)}}{\text{(static pressure)}}$$

From its definition, it can be seen that I is dimensionless. Static probe pressures used to calculate influence are obtained by monitoring landfill gas conditions over a period of time prior to beginning gas extraction testing. Mean static pressure can then be used to calculate influence. Diurnal fluctuations in static pressures have been observed; therefore, the pressure data used to calculate influence should always have been obtained at approximately the same time of day. Experience has shown that static pressures are usually most reliable and at their maximum values during the afternoon hours.

For a given flow rate and at a given depth below the surface of the landfill, influence is expressed as a function of the radial distance (r) from the well being pumped. Typically, influence is reported to be linearly dependent on the logarithm of radial distance as follows:

$$I = A + B \ln r \tag{33}$$

where,

I = influence (as defined above)

r = radial distance from the well being pumped

A = a positive constant

B = a negative constant

A and B are considered constant only if depth and extraction rates remain unchanged. The values of these constants are calculated from the extraction test data. For one withdrawal rate, I versus ln r is plotted for all monitoring probes located at the same depth. A straight line fit through the plotted data is obtained by linear regression analysis. The line's Y-intercept is A; slope of the line is B.

The validity of applying equation (33) to withdrawal of gas from landfills remains to be verified. The equation is analogous to approaches applied to pumping of ground water from an aquifer, where the piezometric surface of the ground water is taken to be linearly dependent on the logarithm of distance from the well [53]. However, for fluid withdrawals from a reservoir, it has been noted that pressure is linearly dependent on ln r for liquids, whereas the square of pressure is linearly related to ln r for gas [22]. Both of these somewhat analogous situations differ from the case of landfill gas recovery in that the withdrawal of liquid or gas from a reservoir does not account for continuous generation of the fluid within the reservoir, as a proper landfill gas extraction model should.

One problem with the expression $I=A+B \ln r$ is that $I=0$ at some finite distance from the well (i.e., at $r=e^{-A/B}$), and this distance is usually referred to as the "radius of influence". Actually, the influence of gas withdrawal extends to the limits of the landfill (and beyond); however, with the instrumentation available, its measurement becomes impossible at some finite distance. The concept of a "radius of influence" has been defined as the distance from the well at which (1) "there is no pressure effect from gas extraction (pressure gradient is approximately zero)" or (2) a pressure of -3 mm of water is observed [5, 6]. Generally, the distance taken as the "radius of influence" depends on the precision of the instruments used to measure landfill gas pressures and on the effects of diurnal pressure fluctuations.

For a given withdrawal rate, an equation of the form $I=A+B \ln r$ can be derived for each depth at which monitoring probes are located. Using these expressions, the radial distance at which a selected value of influence occurs can be calculated for each depth. Thus, for a given well and flow rate, coordinates in terms of depth and radial distance can be attained for points of equal influence. Connecting these points approximates a curve of iso-influence, analogous to elevation contours on a topographic map. Approximation of a surface of iso-influence is attained by symmetrically rotating an imaginary curve of iso-influence around the withdrawal well, assuming complete homogeneity of the refuse. The fictitious, "zero influence" iso-surface is commonly used to calculate "production rate". The "zero" curve of iso-influence is often represented as a straight line to simplify computations, and it has been represented as a straight, vertical line.

Estimation of Gas Production Rate

A completely satisfactory method of estimating a landfill's gas production rate from extraction testing that draws only from a confined radius of influence has not yet been demonstrated. An accurate, theoretical mass balance on the landfill gas remains to be developed and would prove invaluable in making such estimates. The mass-balance could account for refuse characteristics (e.g., permeability) and continuous gas production; the gas composition and extraction rate and the observed internal landfill pressure (or influence) distribution could be related to available equations for convective and diffusive gas flow, accounting for recovery efficiency and loss of gas to the atmosphere as a function of distance from extraction well, landfill geometry, cover conditions, etc. An attempt to correlate these many factors can quickly become very complex, and an effort should be made to keep the model as simple as practicable.

The approaches which have been taken to arrive at quantitative estimates of gas production rate from landfill extraction testing are summarized below.

A simplistic mass-balance approach utilizing the concept of "radius of influence" and implicitly assuming 100% recovery efficiency was used to estimate the gas production rate in the Palos Verdes work. Steady-state conditions and a gas withdrawal rate equal to the gas production rate were assumed. It was assumed that no gas is drawn to the well from a distance greater than the "radius of influence", and that all gas produced within the "radius of influence" is recovered. For these assumptions, flow rate across an imaginary cylindrical surface concentric with the well can be expressed as follows:

$$Q = \pi(R^2 - r^2)t\ D\ F_g \qquad (34)$$

where,

Q = magnitude of the gas flow rate across the imaginary cylindrical surface

R = "radius of influence"

r = radius of the imaginary cylindrical surface

t = refuse thickness

D = refuse density

F_g = gas production rate per unit mass of refuse

Note that $\pi(R^2 - r^2)t$ represents the volume of refuse contained between the surface at the "radius of influence" and the surface crossed by the flow rate, Q, at radius r.

The assumption of steady-state conditions greatly simplifies the approach and is probably appropriate where gas withdrawals of short duration are concerned, in spite of the fact that F_g may fluctuate and the mean F_g will change gradually over a period of years. The other assumptions are suspect; however, they do make the problem more manageable. In reality, some gas will inevitably be drawn across the fictitious "radius of influence", and not all gases produced within the "radius of influence" will be recovered; gas withdrawal rate will never equal the gas production rate because some gases will always escape to the atmosphere. As radial distance from the well increases, a decreasing percentage of the gas produced at that distance will be drawn to the well (i.e., gas recovery efficiency will decrease). Nevertheless, a similar model is most often used to estimate the gas production rate, F_g, of landfills.

Typically, gas withdrawal tests are performed at varying extraction rates until attainment of the maximum extraction rate that minimizes air intrusion and yields consistent quality gas of adequate heating value. The premise is that this gas withdrawal rate is equal to the rate at which gas is being produced within a refuse volume bounded by the "zero-influence" iso-surface (defined by the "radius of influence"). If we refer to this "optimal flow rate" as Q_w* and to the observed "radius of influence" during gas extraction at the rate Q_w* as $R*$, then, neglecting the radius of the well in comparison to $R*$, an expression for F_g is written as follows:

$$F_g = \frac{Q_w^*}{\pi \, (R^*)^2 t \, D}$$

Note that this approach assumes a vertical iso-surface of "zero influence". This is the single most common method by which "gas production rate" has been estimated.

The estimate of F_g might be more appropriately called the "gas recovery rate", since not all gas produced will appear at the recovery well. The rate, F_g, reflects both gas production and recoverability, and it is a useful tool in planning and designing gas recovery systems. But it must be remembered that the value of F_g for a given landfill is not an invariable estimate; its value may depend on well depth and spacing, and on the capabilities of the instrumentation used and the judgement exercised in estimating R*. For example, it is likely that an estimate of F_g based on simultaneous extraction from a number of closely spaced wells would be greater than an estimate based on withdrawal from a single well, because greater recovery efficiency would be attained at points between wells.

The Scholl Canyon work employed a modification of this method for estimating "gas production" in an attempt to reduce inaccuracies inherent in the assumption that all gas produced within the "zero influence" iso-surface is being extracted at the test well [12]. The modified approach assumed that at some radial distance, r, from the withdrawal well, the fraction of gas produced which reaches the well is equal to the influence at that distance. Geometry of iso-surfaces was developed for a number of influence values. The volumes and sub-masses of refuse contained between successive iso-surfaces were computed. For purposes of estimating gas production rate, fractional "contributory mass" of refuse was taken to be equal to the sub-mass times the mean influence between successive iso-surfaces. Extraction rate divided by the summation of the fractional contributory weights gives an estimate of the gas production rate per unit mass of refuse. This method of estimating gas production rate may be used for well extraction rates other than the "optimal" rate.

The Scholl Canyon approach may be an improvement over the more simple method of estimating production rate in that an attempt is made to account for decreasing recovery efficiency with increasing distance from the withdrawal well, and the iso-surfaces are not assumed to be vertical. But the assumption that recovery efficiency is equal to influence should be investigated further. The Scholl Canyon approach assumes that no gas is drawn across the fictitious "zero-influence" iso-surface, as does the Palos Verdes work (however, for the Scholl Canyon test results, it was found that there was a measurable influence to the limits of refuse). Since the Scholl Canyon approach does not assume 100% recovery efficiency within the radius of influence, this method will result in a more conservative estimate of production rate than will the Palos Verdes approach.

The methods outlined above for estimating gas production rate can be applied to the results of either short- or long-term extraction testing; however, short-term testing results may be affected by gas stored within the landfill voids. When short-term testing results show that gas of desirable quality can be withdrawn at reasonable rates without an excessive influence on internal landfill pressures, long-term testing may be conducted to provide greater confidence in gas production rate estimation.

A minimum estimate of the sustainable gas recovery rate can be made from long-term test data by assuming that after a volume of gas equivalent to the landfill void volume has been extracted, all gas extracted is "newly produced gas" (i.e., gas generated during the extraction testing). The "newly produced gas" volume extracted, divided by the duration of extraction testing, places a lower limit on the current recovery rate that could be sustained. Of course, until a deterioration in gas quality is observed, there is no guarantee that the extraction rate employed equals or exceeds the optimum recovery rate for the test recovery system. Also, the spacing, configuration, and number of recovery wells would affect the system's "optimum" recovery rate. Unless the "optimum" system extraction rate has been reached, the minimum sustainable recovery rate so calculated is a conservative estimate. Assuming a constant methane concentration for the extracted gas, the following equations can be used to calculate an estimate of the minimum sustainable methane recovery rate:

$$Q = \frac{(X-V)C}{f\ P} \qquad (36)$$

$$r = \frac{(X-V)C}{f\ M\ P}$$

where,

- Q = lower limit on sustainable methane recovery rate for the entire site
- r = lower limit on sustainable methane recovery rate per unit mass of refuse
- X = total volume of gas extracted
- V = landfill void volume
- C = methane concentration (percent ÷ 100)
- P = duration of extraction testing
- M = refuse mass
- f = fraction of in-place refuse from which the gas was extracted

Note that f is a gross parameter accounting for the relative location in the landfill of the extraction wells and the loss of landfill gas to the atmosphere. Judgment must be exercised in selecting probable values for f, and normally a range of values should be employed, resulting in a probable range for the minimum estimate of sustainable recovery rate. Ideally, a long-term testing program would use extraction wells spaced out over the entire fill and it could be assumed that $f \simeq 1$, but cost considerations usually make complete well coverage impractical.

Refuse Permeability

Darcy's law was employed in the Palos Verdes report to describe landfill gas flow. The equation was used in the form of equation (30), which is repeated below.

$$\upsilon_r = \frac{-k_s}{\mu} \frac{dp}{dr}$$

where,

υ_r = apparent gas velocity at r in a direction toward the well

k_s = specific or "apparent" intrinsic permeability of the medium

μ = absolute viscosity of the gas

r = radial distance from the recovery well

p = pressure at distance r

The apparent gas velocity as a function of r was then expressed as follows:

$$\upsilon_r = \frac{Q}{A} = \frac{Q}{2 \pi r h}$$

where,

Q = volumetric flow rate

h = depth of well

These equations were combined and rearranged, permitting integration of the resulting differential equation, as follows:

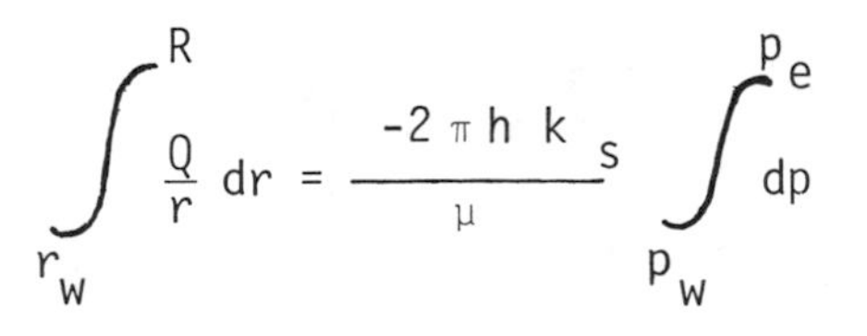

$$\int_{r_w}^{R} \frac{Q}{r} dr = \frac{-2 \pi h k_s}{\mu} \int_{p_w}^{p_e} dp$$

where,

r_w = radius of the well

R = "radius of influence" of the well

p_w = pressure at the well/refuse interface

p_e = pressure at the "radius of influence" of the well

Here, h, k_s and μ are assumed to be constant and have been taken outside the integral. However, Q has been left inside the integral because it is a function of r. Gas production takes place throughout the refuse. Consider two values of r, r_1 and r_2, where $r_2 > r_1$, and imagine concentric cylindrical surfaces at distances r_1 and r_2 from a gas withdrawal well. Flow rate across the surface at r_1 must be greater than that across the surface at r_2 because of the additional contribution of gas from the continuous production taking place in the volume of refuse between the two surfaces. Therefore, proper evaluation of the integral on the left side of the equation would require that Q be expressed as a function of r.

In the Palos Verdes report, Q was treated as a constant and was taken outside the integral, resulting in the following expression for k_s:

$$k_s = \frac{Q_w \mu \ln(\frac{R}{rw})}{2\pi h \Delta p} \qquad (37)$$

where,

$$\Delta p = p_e - p_w$$

Note that the negative sign has been omitted from equation (37); as equation (37) is written, the magnitude of the volumetric well flow rate, Q_w, should be used without regard to sign convention for direction of flow.

This approach may be appropriate for withdrawals from a reservoir where no gas production is taking place, but applying it to an active landfill results in neglecting gas production. Since the same flow rate was assumed to exist across all "cylindrical surfaces" regardless of their distance from the well, the entire gas flow was taken to originate at the "radius of influence". Note that the assumptions applied here are diametrically opposed to those applied in the Palos Verdes method of production rate estimation, where it was assumed that all gas extracted originates within the "radius of influence". The shortcomings of this method of estimating refuse permeability again emphasize the need for development of a reasonably accurate mass-balance on the gas.

From gas extraction tests conducted at the Palos Verdes landfill, it was concluded that the "radius of influence" was 76 m for a 34 m deep well at a flow rate of 150 L/sec with an applied suction of 34.4 cm of water below atmosphere. These values and an estimate of 0.013 centipoise were substituted into the above equation, and k_s was calculated as 20 darcys.

Refuse permeability was also estimated in the Sheldon Arleta report [26]. In that work, the Kozeny equation was used to describe gas flow, and, as in the Palos Verdes report, gas production was neglected. At Sheldon Arleta it was reported that the "apparent permeability coefficient" increased with an increase in the vacuum applied at the withdrawal well. This is not a surprising result in that the greater applied vacuum should have permitted more efficient recovery of gases produced. For the "deep" wells, with an applied vacuum of 30 cm of water, the "radius of influence" was estimated to be 76 m and the "apparent permeability coefficient" was estimated as 0.0012 cm/sec. The value of γ for the composite gas was reported to be 1.07 kg/m^3. Assuming that $\mu = 1.15 \times 10^{-5} Nsec/m^2$ for the composite gas, the "apparent permeability of the refuse, k_s, can be calculated as follows (see equation 29):

$$k_s = \frac{\mu}{\gamma} K \; \frac{1.15\text{x}10^{-5} \frac{\text{Nsec}}{\text{m}^2}}{1.07 \frac{\text{kg}}{\text{m}^3}} \times 1.2\text{x}10^{-5} \frac{\text{m}}{\text{sec}}$$

$$\times 1\text{x}10^4 \frac{\text{cm}^2}{\text{m}^2} \times \frac{1 \text{ kg}}{9.8 \text{ N}} \times \frac{1 \text{ darcy}}{9.85\text{x}10^{-9}\text{cm}^2}$$

$$k_s = 13.4 \text{ darcys}$$

This value is reasonably close to the estimate of 20 darcys reported for Palos Verdes under similar pumping conditions.

Design Criteria

In the Palos Verdes report, an expression for the magnitude of gas velocity toward the well, v , was written by dividing equation (34) for Q by the area of the imaginary cylindrical surface of radius r, as follows:

$$v_r = \frac{Q}{A} = \frac{\pi(R^2-r^2)hD\,F_g}{2\,\pi\,r\,h} = \frac{(R^2-r^2)D\,F_g}{2r} \qquad (38)$$

This approach assumes that landfill conditions are uniform, that F_g is the true gas production rate, that all gas produced within R is collected, and that no gas is drawn across the imaginary surface at R. Actually, only a fraction of the gas produced at some distance r from the well would be collected, and this fraction would decrease as r increases.

As typically calculated, the value of F_g accounts for efficiency of recovery under the specific conditions of the gas extraction testing, and when the contributing volume of refuse surrounding the well is considered in bulk (i.e., F_g reflects the "average" or overall recovery efficiency). In this context, F_g does not represent the true gas production rate, but rather, it represents the maximum recovery rate for the system tested. Introducing F_g into the equation for v_r is bound to create some inaccuracy where F_g is taken to be a constant, because recovery efficiency will decrease as radial distance from the well increases.

The Palos Verdes approach proceeds, equating the above expression for v_r with Darcy's law for v_r (equation 30), as follows:

$$\frac{-(R^2 - r^2) D F_g}{2r} = -\frac{k_s}{\mu}\frac{dp}{dr}$$

(Note that the sign convention for velocity direction toward the well has been included here.)

$$\int_{p_w}^{pe} dp = \frac{\mu D F_g}{2 k_s} \int_{r_w}^{R} \left(\frac{R^2 - r^2}{r}\right) dr$$

Evaluation of the above integrals gives the following expression:

$$\Delta p = \frac{\mu F_g D}{2 k_s}\left[R^2 \ln\left(\frac{R}{r_w}\right) + \frac{r_w^2}{2} - \frac{R^2}{2}\right] \qquad (39)$$

An improvement on this approach could be made if F_g were treated as a function of r rather than as a constant, or if an expression for recovery efficiency as a function of r were introduced into the equation and F_g were treated as a constant value (throughout the landfill volume) of the actual, composite gas production rate.

Equations (34) and (39) were used in the Palos Verdes report to correlate R with Q_w or Δp for various assumed values of F_g. For example, equation (34) could be used to estimate the well flow rate, Q_w, that would be attained for a given value of R, or to estimate R for the desired well flow rate; this information is useful in determining inter-well spacing when laying out the recovery well system. Equation (39) could be used to help size the blower unit which applies the suction to the well field or to determine well spacing given a practical value of Δp. Note that the value of k_s calculated from equation (37) may not be appropriate for use in equation (39), as these equations were derived from contradictory assumptions. Anyone using these equations should keep in mind the limitations inherent in the simplifying assumptions used in their development. In particular, since landfill conditions are never homogeneous, accurate modeling of the system is difficult. Judgment, experience, and experimentation are essential to rational, flexible design of a landfill gas recovery system.

Recovery System Management

The design and operation of a landfill gas recovery system is still as much an art as it is a science. In conducting a gas extraction test program and interpreting the resulting data, common sense, experience, judgment, and even intuition are essential assets. For example, experience with gas recovery projects at numerous landfills will enable one to assess generally the gas production potential for a particular landfill from such information as refuse appearance or composition, refuse tonnage, age, and thickness, conditions of landfill cover and surrounding soils, and other available background data. Short of extraction testing, quantitative methods for this assessment have not been developed. Judgment based on experience is critical to a preliminary estimation of methane recovery potential.

Similarly, experience gained during extraction testing from a particular landfill should form a basis for decisions related to management of an eventual, full-recovery system. For example, extraction testing can give an indication of the best gas producing wells and areas of the landfill, initial extraction rates to employ at particular recovery wells, and other operational criteria.

METHODS OF RECOVERING LANDFILL GAS

GENERAL CONSIDERATIONS

Any recovery system for methane use should be designed and operated in a manner which protects the long-term viability of the landfill as a methane generator. Conditions favorable to methane production should be maintained, as well as possible, throughout the landfill. For example, a strategy of minimizing air intrusion into the fill should be employed. "Overstressing" a landfill by extracting gas at too great a rate will cause substantial air intrusion, resulting in the establishment of aerobic regions within the landfill and a reduction of the gas recovery rate and of the total gas volume to be captured.

INDUCED EXHAUST WELL SYSTEMS

General Description

The induced exhaust well system is the only method of gas recovery that has been used in cases where beneficial use of the captured gas has been intended. This system has also often been employed when control of lateral gas migration was the objective. In many circumstances, it is desirable to design a system which will both control gas migration and efficiently recover methane for beneficial use.

The strategy of minimizing air intrusion is important from the standpoint of a recovery and use objective, rather than for gas control. Air intrusion can be minimized by using closely spaced extraction wells operating at relatively low flow rates, or by providing an effective gas containment system around the refuse (e.g., synthetic membrane or tight soil). Of course, there is an economic tradeoff, and a point may be reached where the cost of improved gas recovery facilities outweighs the potential benefit to be gained by improving recovery efficiency.

Also, improvement of the gas containment systems is normally impossible (except for improving cover conditions) in completed portions of a landfill.

Design criteria for exhaust wells, including their depths, collection intervals, inter-well spacings, and setback distances from the perimeter, are determined from site geometry and other background information. The results of previous extraction testing and the program objectives (i.e., migration control and/or recovery and use) also help determine exhaust well design criteria.

An induced exhaust well for the recovery of landfill gas typically consists of a perforated pipe casing placed in a hole drilled in the refuse, backfilled around the perforations with gravel, and sealed off against the inflow of atmospheric air around the outside of the well casing. Recent innovations in well design include installation of driven perforated pipe and placement of perforated pipe in slightly undersized drilled holes. However, much remains to be learned about the relative merits of the various well designs. A suction is applied to each well casing, creating a pressure gradient within the vicinity of each well which decreases in the direction of the well. The gas is gathered to the well by the predominantly convective flow and is then removed at the well head. Each well head is normally equipped with a butterfly valve for flow rate control, permitting flexibility in system operation.

The gas withdrawn at each well is collected to a central point by means of a pipe network referred to here as the gas collection header. A blower/motor unit is normally the source of the applied suction and the central point to which gas is collected. Gas transmission or use may require pressures greater than the 25 to 38 cm of water available from a blower, in which case a gas compressor would follow the blower. Disposal of excess gas to the atmosphere will usually require a gas burner for control of malodours.

The following discussions include more details on the various components of an induced exhaust well system for landfill gas recovery.

Well Construction

Drilling

Well construction begins with the drilling of a borehole in the refuse. Borehole diameters in the range of 60 to 90 cm are commonly used and have been found suitable; however, smaller diameter holes may be adequate in some instances. Larger diameter boreholes provide greater surface area at the refuse/gravel interface and, logically, should require a smaller applied suction to attain a given flow rate than do smaller drill holes since the gravel backfill is more permeable than the refuse removed.

Well depths ranging from 50 to 90% of the refuse thickness are common. The existence of a permanent leachate level above the base of refuse is a lower limit on well depth. Typically, collection intervals (i.e., well length along which the pipe casing is perforated or the borehole is backfilled gravel) vary from the lower one third to the lower three quarters of the well depth. In any event, the upper extent of the collection interval should be far enough below the ground surface to preclude substantial air intrusion.

Drilling in refuse is usually accomplished with a truck-mounted continuous flight auger rig. Rates of progress are often slow due to difficulties encountered in drilling refuse. For example, construction debris (e.g., lumber and concrete) and large metal objects such as automobile bodies and household appliances may cause delays, and holes drilled in poorly compacted refuse may cave in, requiring the removal of additional refuse. For the drilling of 60 to 90 cm boreholes the drill rigs should have an average rate of progress ranging from about 3.5 to 10 m per hour depending on fill conditions.

Materials

Various materials have been used for the well casing and associated fittings. Although polyvinyl chloride (PVC) pipe is the most common material, fiberglass, polyethylene, and steel also have been used. PVC has been the most popular choice because of its relatively low cost, and its generally satisfactory performance. On the other hand, PVC pipe may be unacceptable where gas temperatures are high, and PVC is known to deteriorate rapidly when exposed to ultraviolet radiation (present in sunlight).

Pipe casings are sized on the basis of the expected gas flow rate and the permissible pressure loss from the well head to the collection interval. Casings with diameters of 8 to 15 cm are common. Often, two casing diameters are used, with alternating casing lengths of different diameters which permit telescoping of the casing to accommodate landfill settlement. Alternating pipe lengths of different diameters are joined by slip joints which permit the pipes to slide together but prevent their pulling apart beyond some fixed point. These telescoping joints are used mainly where refuse thicknesses are great; their performance remains to be documented.

Figures 10 and 11 show typical telescoping gas well details.

Along that length of the exhaust well from slightly below to slightly above the well casing's perforated interval, the borehole is backfilled with gravel to facilitate the flow of gas from the refuse to the perforations. Commonly, gravel of a size less than 4 cm is used. A concrete plug 60 to 90 cm thick is normally poured around the casing, above the gravel backfill, as a seal against the inflow of air from the surface. A relatively dry concrete mixture should be used to prevent plugging of the gravel-filled gas collection interval of the well. Above the concrete plug, the borehole is backfilled around the well casing with fine-grained soil which acts as a further seal against air intrusion.

Each well head is normally connected to the gas collection header by means of flexible hose, which allows for thermal expansion/contraction and differential settlement. A vault box over the well head is normally provided to limit accessibility to the flow control valve.

Perforations

Several well casing perforation patterns and methods have been used. The primary requirements of the perforations are that they remain open, that they do not require excessive pressure losses to draw the gas through them, and that they do not unduly weaken the pipe. Generally, the casings have been perforated using either a drill or a saw. Drill diameters of 1.3 to 2.5 cm are commonly used, and four or more holes may be drilled in one ring around the pipe casing. Individual rings of perforations might be spaced 10 to 30 cm apart. Weakening of the pipe casing can be minimized by radially staggering alternate rings of perforations. Sawed slots of 3 to 6 mm in width have also been used as perforations. The slots are commonly cut to a depth of 1/6 to 1/3 the outside diameter of the pipe casing. Relatively shallow perforations might be sawed in pairs with perforations located opposite one another. Alternate pairs could then be staggered 90°, radially, to maintain pipe strength. Deeper perforations might be sawed individually, with adjacent perforations staggered 180°, radially. The longitudinal spacing of sawed perforations is usually within the range of 8 to 20 cm.

Well Spacing and Layout

Well spacings are normally determined using the "radius of influence" concept and an approach similar to equations (31) or (32). Assuming typical values of 650 kg/m^3 for refuse density, 7 mL/kg/day for the methane recovery rate per unit mass

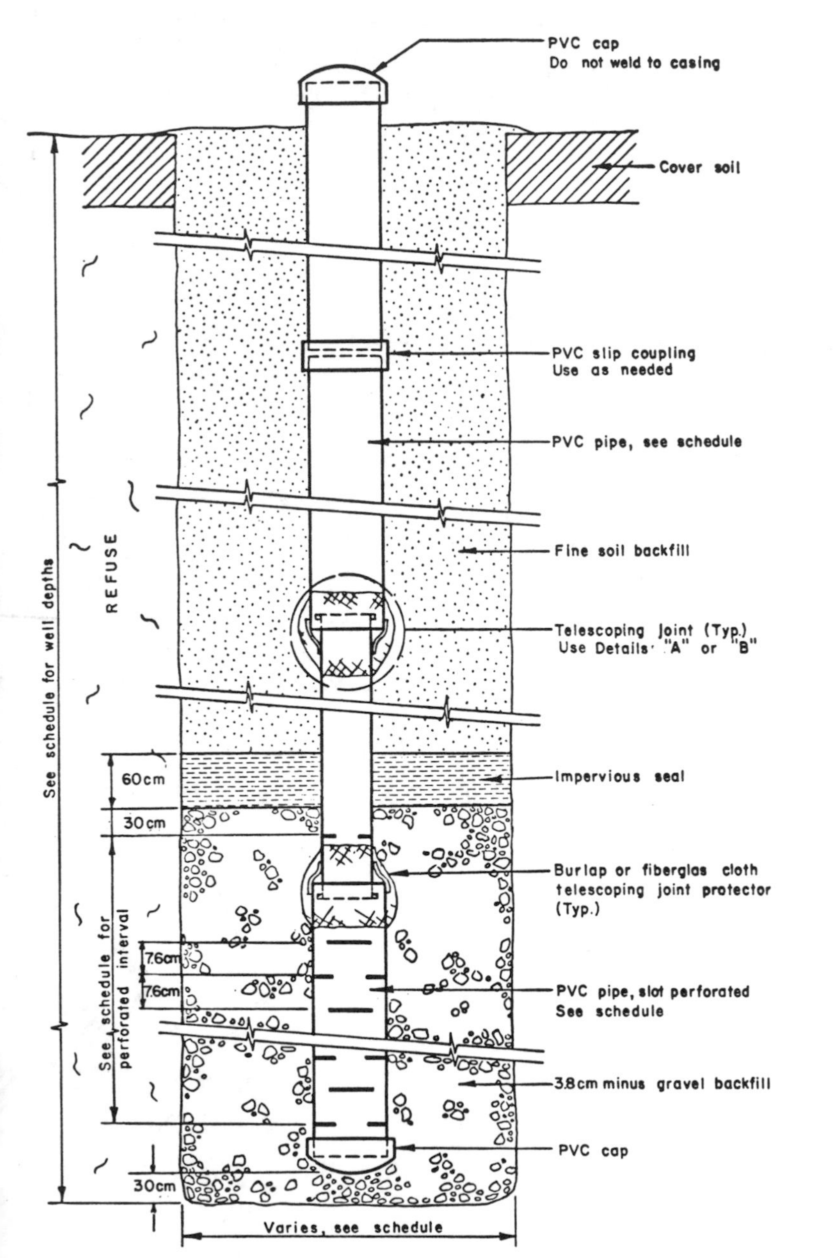

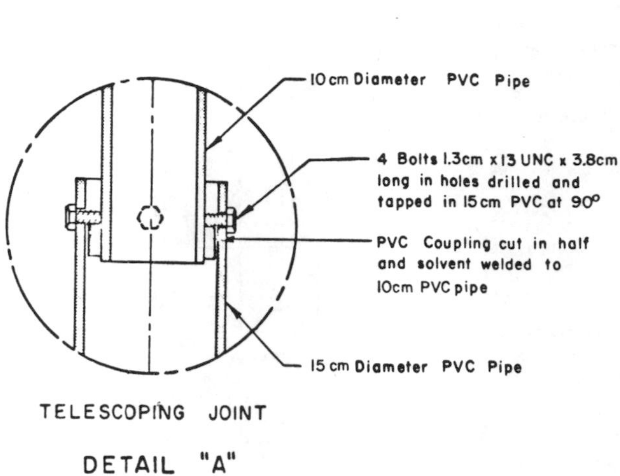

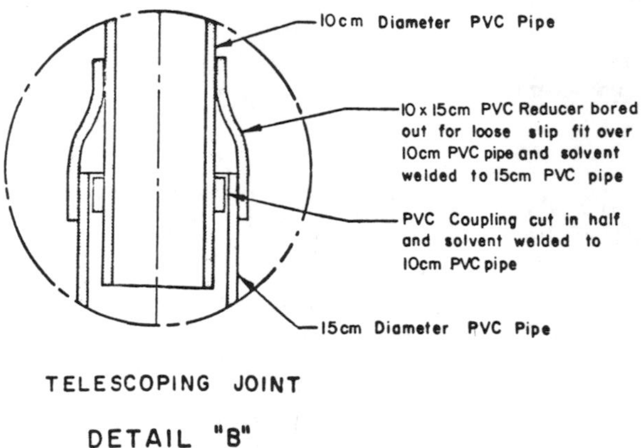

Not to scale

WELL SCHEDULE

WELL NUMBER	WELL DEPTH (m)	WELL DIAMETER (cm)	PERFORATED INTERVAL (m)	WELL HEAD CASING SIZE (cm)	BOTTOM SECTION CASING SIZE (cm)	NUMBER OF TELESCOPING JOINTS

NOTES

1. Perforations are typically .3cm wide slots cut to a depth of 1/6 the outside diameter of the casing. Slots are cut in pairs located 180° from one another; adjacent pairs are staggered 90° radially.
2. Solvent weld all PVC couplings and bottom caps.
3. Deep wells may have additional telescoping sections.
4. More than one pipe length may be perforated.
5. All PVC pipe and fittings are Schedule 40.

Figure 10. Typical design of gas recovery well (Version 1).

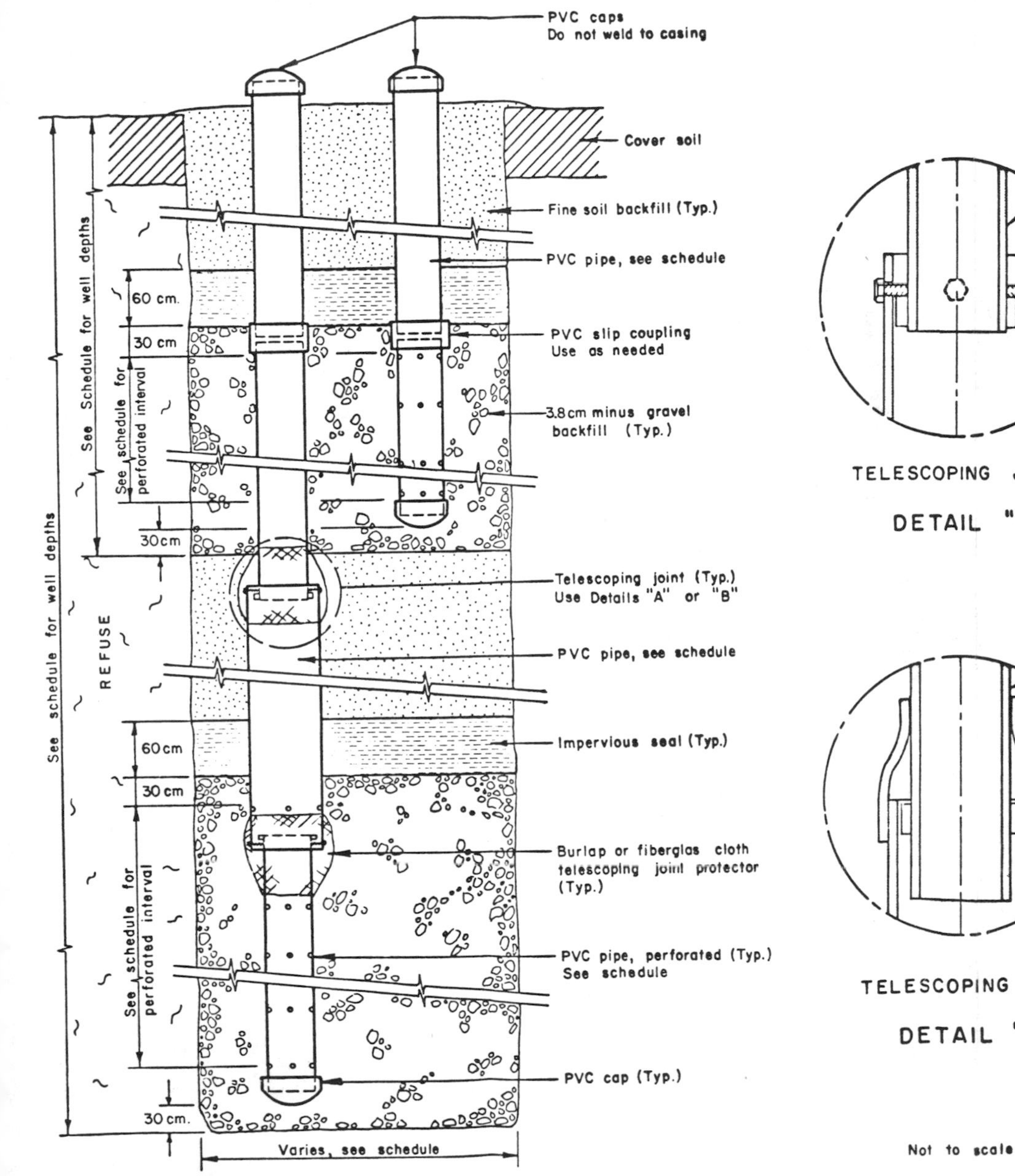

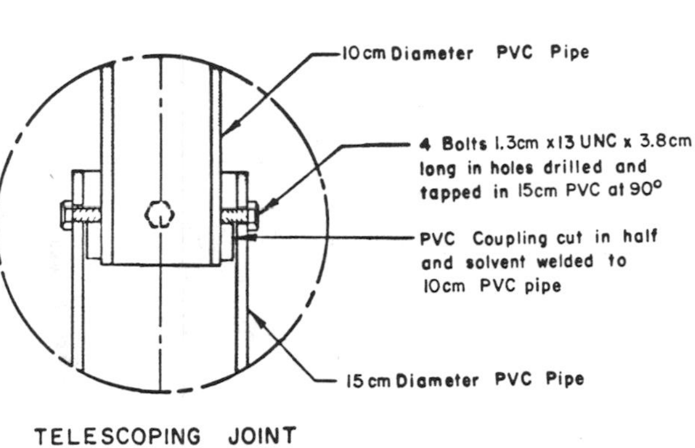

TELESCOPING JOINT

DETAIL "A"

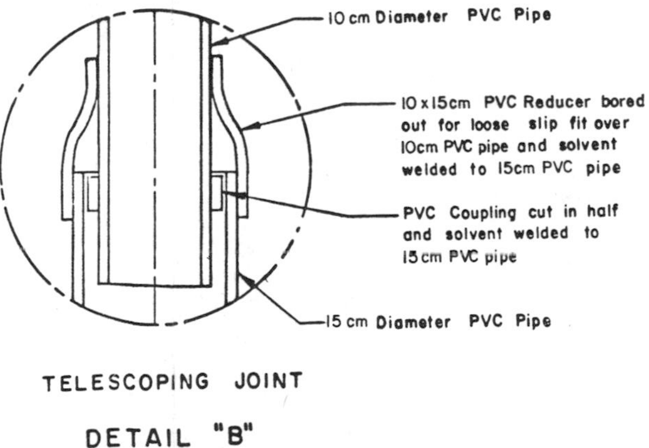

TELESCOPING JOINT

DETAIL "B"

Not to scale

WELL SCHEDULE

WELL NUMBER	WELL DEPTH (m)	WELL DIAMETER (cm)	PERFORATED INTERVAL (m)	WELL HEAD CASING SIZE (cm)	BOTTOM SECTION CASING SIZE (cm)	NUMBER OF TELE-SCOPING JOINTS

NOTES

1. Perforations are typically 1.9cm holes spaced 15cm longitudinally and 90° radially.
2. Solvent weld all PVC couplings and bottom caps.
3. Deep wells may have additional telescoping sections.
4. More than one pipe length may be perforated.
5. All PVC pipe and fittings are Schedule 40.

Figure 11. Typical design of gas recovery well (Version 2).

of refuse, and a 50% methane concentration, equation (32) could be used to correlate well flow rate, Q_W, with "radius of influence", R, given the refuse thickness, t. For example, if gas is extracted at a rate of 50 L/sec from a 30 m deep landfill, a "radius of influence" of about 70 m could be expected. The gas extraction rate required to establish a "radius of influence" of about 50 m would be estimated at 25 L/sec. Inter-well spacings should be less than or equal to twice the estimated "radius of influence".

Some "overlap" of influence is desirable for the perimeter wells of a system designed for control of gas migration to insure that effective control is obtained at points between wells along the landfill boundary. Of course, gas extraction rate and "radius of influence" are dependent on one another, and individual well flow rates can be adjusted after the recovery system is in operation to provide effective migration control and/or efficient methane recovery. In employing this approach, it should be remembered that the "radius of influence" is not a concept that can be taken literally; it is merely a convenient tool for simplifying engineering calculations.

Once a desired inter-well spacing and well setback distance from the landfill boundary have been established, recovery wells are established in a pattern providing efficient recovery of gas. Given no constraints, locating exhaust wells at the vertices of equilateral triangles (the sides of which are equal to the desired inter-well spacing) provides the most complete coverage of a landfill's area. It is usually not feasible to lay out an entire well field in the equilateral triangle pattern because of irregularities in the geometry of the landfill boundary. It is also difficult to maintain the equilateral triangle pattern because smaller inter-well spacings are often used for wells nearest the landfill perimeter than are used for interior wells. Perimeter wells must be relatively closely spaced and operate at low flow rates to minimize air intrusion across the refuse/soil interface. Interior wells are less susceptible to air intrusion and can utilize greater inter-well spacings and higher well flow rates, minimizing the number of wells required and decreasing construction costs.

A recovery well layout design should take into account the anticipated end use of the land. This type of planning is particularly relevant when layout of the gas collection header is combined with the well layout. For example, well heads and collection header piping should usually be laid below final grade if recreational use is anticipated. Wells and collection header should normally not be located where structures are to be constructed within the gas recovery project life.

Gas Collection Header

Materials

Polyvinyl chloride (PVC) pipe has been most widely used in gas collection headers, primarily because of its low cost and easy installation. However, as indicated above, PVC should not be used in piping exposed to long-term solar radiation because the material is known to deteriorate under ultra-violet radiation. High density polyethylene pipe has also been utilized for gas collection header pipe. Recently fiberglass-reinforced resin pipe impregnated with a reflective material to resist attack under ultra-violet light has been used for above-ground installations.

Pipe Protection

Gas collection header piping usually requires protection from damage by exposure to the elements, equipment traffic, and vandalism. Soil cover can usually be expected to provide the best practical means of pipe protection. The choice is whether to lay the pipe on the surface and mound soil over it or to lay the pipe in a trench and backfill with soil. The required performance of the pipe protection, the end use planned for the site, and the relative cost of the alternatives should be weighed in making a decision. For example, consider a landfill located in a warm climate: if there is no end use planned for this site, a shallow soil cover over the pipe is sufficient to protect it from vandals and ultra-violet light, provided equipment traffic is not a problem. But trenching may offer the best alternative, for example, if there is a planned end use requiring a flat surface or if the landfill is located in a climate where the header pipes must be protected from freezing of condensate during the winter months.

Pipe Slopes and Condensate Drains

Gas recovered from a landfill normally has been found to be saturated with moisture. During collection in the header system, the gas undergoes an expansion and temperature decline, and some water condenses (falls out of the gas), accumulating in the bottom of the header lines. If the condensate is allowed to accumulate in low spots of the line, the header pipe may become partially or completely blocked. To avoid the line blockage problem, condensate drains should be constructed at regular intervals along the header line, and the line should be always sloped toward a condensate drain.

Gas collection piping is usually sized on the basis of gas flow requirements. Given pipe sizes and an assumed or calculated rate of condensate formation, the header pipe slopes and condensate drain spacing can be established to keep only a negligible portion of the pipe cross-sectional area occupied by condensate. The required pipe slopes can be determined by methods analogous to those used in sewer design. For example, the Manning or Hazen and Williams equations could be used to evaluate the hydraulic characteristics of the pipe system. Adequate field inspection and supervision should be available during construction of the header line to insure that the lines are sloped according to specifications. Buried header lines may present difficult problems if pipe slopes are incorrect, and differential settlement of the landfill may cause adverse slopes which require remedial measures.

Condensate drains should be located at all low spots and at more or less regular intervals along the gas collection header line. Typical spacings of the drains might range from 60 to 210 m. Shorter spacings are preferred because the drains are relatively low cost items and the chances of major blockage due to differential settlement diminish as the number of condensate drains increases.

A condensate drain basically consists of a small diameter (e.g., 2.5 cm) pipe connected to the header line by a "T" joint. The pipe extends downward from the header line at a low spot in the collection pipe; the accumulated condensate drains out of the header line through this "dripleg". Since there is suction applied to the header line, means must be provided for preventing air flow through the dripleg and into the header line. Typically, the lower end of the dripleg is immersed in a reservoir of water or condensate. The length of dripleg, from its upper opening in the header line to its lower opening immersed in the reservoir, must be greater than the greatest anticipated applied pressure, relative to atmospheric pressure (expressed as a height of water column).

The distance from the reservoir overflow to the top of the dripleg must also be greater than the height equivalent of the greatest anticipated relative pressure to be applied, so that excess condensate will flow from the reservoir and not be sucked back up into the header line. Above any openings in the lower portion of the dripleg, a reservoir volume must be maintained sufficient to fill the dripleg to a height above the reservoir surface equivalent to the greatest relative pressure to be applied.

Excess condensate is most often percolated back into the refuse, from whence it came. The condensate reservoir is typically set in a gravel-filled excavation, and the overflow from the reservoir is applied to the gravel.

Header Pipe Sizes and Layout

The absolute pressure change experienced in typical landfill gas collection headers would almost always be less than 5% (5% of atmospheric pressure is equivalent to about 38 mm of mercury), and, for all practical purposes, gas flow in the header lines can be considered as incompressible. If pipe slopes and condensate drains are well designed and properly constructed, the pipe cross-sectional area which is occupied by condensate can be neglected without materially affecting the sizing of header piping on the basis of gas flow.

Header pipes are sized on the basis of flow rate and permissible pressure loss in the header line. Flow equations analogous to those used for design of air and water distribution systems can be used to correlate pressure loss with gas flow rate, pipe diameter, pipe length, and pipe material. Using a commonly employed pipe-friction equation and assuming a header pipe of circular cross-section, pressure loss would be directly proportional to (1) a friction factor, (2) pipe length, and (3) the square of mean gas velocity; and inversely proportional to (1) pipe diameter, and (2) twice the acceleration of gravity.

The value of the friction factor can be determined, for example, from a Moody chart. The value of the friction factor is a function of the relative roughness of the pipe (i.e., absolute roughness divided by pipe diameter, a dimensionless number) and the Reynolds number. The Reynolds number is the product of the pipe diameter and the mean gas velocity, divided by the kinematic viscosity of the gas (the Reynolds number is dimensionless). The kinematic viscosity is absolute viscosity divided by gas density, and, for a given gas composition, kinematic viscosity is a function of gas temperature. Methods of solution for the flow of gas in air conditioning and heating networks are in common usage, and standard texts should be consulted for details.

The first step toward designing a gas collection header line is to estimate gas flow rates from the individual recovery wells. For a proposed header layout, flow rates in each segment of the header line can be estimated. Since preliminary flow rate estimates may be inaccurate, a factor of safety should be used to adjust the flow rate upward, especially where control of lateral migration is an objective. If the header line for a gas control system is conservatively sized, the perimeter wells may eventually be operated at higher flow rates than originally anticipated, and additional perimeter wells can be constructed at points along the header line where control is inadequate.

Once the flow distribution has been estimated for a proposed header layout, segments of the header pipe can be sized to keep pressure losses within reasonable limits, as determined by the performance characteristics of available motor/blower units. Numerous header layouts can be proposed, each requiring different lengths of various sizes of pipe. Since the installation cost per metre (linear foot) is less for smaller pipe sizes than it is for larger sizes, there is an economic tradeoff; for example, a proposed layout using a greater total length of header pipe may be less expensive than a layout using a lesser total length of pipe but using more pipe of relatively large diameters. Generally, several proposed header layouts should be investigated in an attempt to minimize construction costs.

Possible Maintenance Problems

Gas collection headers are subject to several potential problems that should be addressed during the design and planning stage. It should be recognized from the outset of any landfill gas recovery project that ongoing expenses for maintenance will be incurred. Since one possible problem with header piping is breakage due to thermal contraction or expansion, header design should provide for flexible connections or layout to avoid such breakage.

Another potential problem with a header system occurs when differential settlement of the fill causes pipe movements resulting in adverse slopes, accumulation of condensate in low spots and partial or complete blockage of gas flow. The condensate blockage problem and the requirement for proper pipe slopes and condensate drains were addressed earlier in this chapter.

In anticipation of problems caused by pipe breakage, condensate blockage, or other header system failure, a regular program of periodic inspection and maintenance should be established. Tools and pipe materials should be readily available for routine repair of the line, and pipe slopes should be checked and adjusted as required.

Above-ground installation of the gas collection header is sometimes possible, and the easy accessibility greatly simplifies pipe maintenance and slope adjustment. If the end use plan for the land, climatic conditions, aesthetic considerations, or security precautions dictate that the header system must be below grade, maintenance becomes more difficult and the cost of maintenance can be expected to increase.

OTHER TECHNIQUES

Combined Recovery/Control Systems

Although recovery and beneficial use of landfill gas can often prove to be a viable energy alternative on the basis of the economic merits of the recovery and use project alone, the concept makes sense in many other situations where there is a hazard potential from the lateral migration of methane toward structures. Properly designed and constructed induced exhaust wells used for gas recovery systems have been shown to effectively control the lateral migration of landfill gas. In addition, completely enclosing the refuse with a liner impermeable to gas would also avoid a lateral migration problem. However, in cases where the economic benefit from the use or sale of the gas does not significantly offset the cost of the control/recovery project, more economical control alternatives may be available.

Shallow Trench Collection Galleries with Induced Exhaust

Although vertical exhaust wells have normally been used for the recovery of landfill gas, as described earlier in this chapter, a network of shallow trenches shows good promise as a feasible method of gas recovery. The shallow trench systems may have special merit where the leachate level is relatively close to the ground surface or where the refuse is thin. The shallow trenches would be backfilled with gravel, and perforated, collection header piping would be laid in the gravel backfill. The trenches and header piping would be laid out and interconnected in network fashion.

An essential element of a workable shallow trench collection gallery is the provision of a good seal above the trench network to preclude air intrusion. In designing a trench collection system, consideration should be given to the alternative of completely covering the site with a synthetic membrane material.

Natural Exhaust

A natural exhaust landfill gas recovery system could be designed to incorporate either vertical wells or shallow trenches. An essential element of any natural exhaust system would be a good seal, approaching a complete enclosure around the refuse, to prevent the escape of landfill gases. For example, the refuse mass could be completely enclosed by a synthetic membrane impermeable to gas. Logically, one should get the greatest methane recovery from a completely enclosed system. Such a system maximizes methane production because anaerobic conditions can be maintained throughout the refuse mass (all other factors affecting methane production assumed to be equal); it also maximizes recovery efficiency because the refuse gas cannot escape to the atmosphere.

UTILIZATION AND PROCESSING OF LANDFILL GAS

In conjunction with the field testing program, a market and background data survey must be undertaken for a site. The survey should determine the potential buyers available in the local area, the energy usage load, the prospective price considerations, and anticipated social, political, and technical obstacles. The survey should also develop the landfill placement history, character and composition of the refuse, moisture history, and site physical characteristics. This survey, coupled with an extensive test program, should provide the basis for prediction whether it is feasible to recover and sell methane from a specific landfill.

If the gas is to be utilized by more than one buyer, the landfill gas seller may be competing with a utility company, thus falling under the jurisdiction of utility control agencies. Thus, entrepeneurs contemplating landfill gas sales and/or purchase should thoroughly investigate sales regulations and restrictions.

UTILIZATION MODES

There are three primary categories of use for methane from landfills: (1) the direct sale of low kilojoule (heating value) gas to industrial customers; (2) utilization of low kilojoule gas as a source of fuel for electrical generation; and (3) the conversion of refuse gas to pipeline standard gas for injection into nearby utility company pipelines.

Other potential uses of landfill gas are:

1. Steam generation as a source of heat or for electrical generation.

2. Conversion of landfill gas to liquefied natural gas (methane).

3. Conversion of landfill gas to methanol (methyl alcohol).

4. Direct combustion for space heating.

5. After removing moisture and particulate matter, injection of low kilojoule gas directly from the landfill into an existing natural gas transmission line.

The sale of low kilojoule gas to nearby industrial customers is generally the most economical and profitable alternative; if there are no regulatory constraints, there are no pricing restrictions. The limiting economic factors for a successful venture are the volume of gas available (including future reserves), the distance from the landfill site to the customer(s), local market price, gas use pressure requirement(s), and compatibility of use as compared with pipeline standard gas or alternative fuels.

Electrical generation is a more costly end use as compared with the direct sale of low kilojoule gas to industrial customers. However, because of the limited low kilojoule gas market for any particular landfill site, one should consider the use of methane gas for electrical generation or some other purpose.

For each individual landfill site, specific market conditions, regulatory constraints, and economics will dictate the preferred end use.

Low Kilojoule Gas

Prior to transmission of low kilojoule gas off-site, some on-site gas processing will be required. The minimum process requirements for low kilojoule gas use consist of free liquids removal and compression of the gas to elevated pressure. Removal of water and other liquids will minimize pipeline entrapment of condensation with its associated problems (corrosion, liquid surge, flow blockage, and unacceptable water content). The compression process provides the necessary vacuum to withdraw gas from the landfill and pressure head to move the gas to its use point.

Designers must select whether to use an electric motor or internal combustion engine (option of dual fuel) for the compressor system. The reliability (on-line time) of combustion engines is between 90 to 95%, whereas the reliability of the electric motor approaches 100%. However, the cost of electricity for the electric motor is higher than the fuel for the combustion engine, because the combustion engine can use landfill gas. In addition, the electricity supply may be interruptible if the utility source has a brown-out or other power outage. Each case must be decided on the merits of its own economy and reliability.

On-Site Generation of Electricity

Gas withdrawn from a landfill can be utilized for on-site electrical generation. Initially, the gas is injected (typically under a pressure of 0.35 to 0.70 bars into an internal combustion engine whose carburation is modified to deliver the proper amount of air (approximately 10%) for combustion. The motor runs a generator to develop electricity. A dual fuel engine can be utilized in the event the thermal heat of combustion falls much below 19 kJ/L; natural gas, propane, or diesel fuels can provide the supplementary fuel supply.

An alternative approach to utilizing landfill gas for on-site electrical generation is to introduce the gas to a gas turbine engine. This method requires a compressor stage ahead of the turbine since the injection pressure is in the range of 10 to 20 bars. A significant drawback of this approach is that the system efficiency is very low because of heat loss. However, the system's efficiency can be measurably improved if the heat loss from the combustion is captured for steam production or other beneficial heat use.

The internal combustion engine-generator system has the following advantages over the gas turbine system: (1) lower capital costs; (2) less compression before injection to the engine; and (3) resulting higher efficiency because of decreased energy requirements for compression.

Upgrading To Pipeline Quality

Landfill gas can be upgraded to yield a variety of heating values (kJ/L) by numerous conventional processes. The cost of upgrading rises dramatically as impurities are separated from the methane. The ultimate product can be essentially pure methane (equivalent to pipeline standard natural gas) with a heating value of about of 38 kJ/L.

Certain users may require partial upgrading to a quality which will meet their need. Although upgrading all the way to pipeline quality is an expensive process, every user of natural gas, as well as the gas transmission companies and utilities, should be able to use this quality gas. Consequently, pipeline quality gas may yield the greatest price premium. In addition, some credit for the by-product CO_2 may be obtained if it can be removed in a relatively pure state and a local CO_2 market exists. Although the process technology to approach or achieve pipeline quality gas is relatively straightforward, "off-the-shelf" technology, some technical problems may be encountered because of variations of trace contaminants found in landfill gas. Therefore, experience in process technology is a necessary prerequisite to successful achievement of pipeline quality gas.

Miscellaneous Uses

Where a commercially viable opportunity does not exist for recovery and sale, landfill gas can be used as a supplement to, or substitute for, miscellaneous on-site fuel requirements. The gas can be gathered by a blower-motor system which can deliver it at a pressure of about 0.34 bars for direct burning. This source of heat energy can be utilized for space heating and other applications where a source of heat is required.

PROCESS TECHNOLOGY

Dehydration for Water Removal

Landfill gas is generated within a temperature range of 27 to 66°C and at a saturation of nearly 100%. The internal pressure within a landfill typically ranges from atmospheric to a pressure ranging from 2.5 to 5 cm of water above atmospheric. As the gas is extracted and moves through the header system to either a motor and blower or to a process station, the temperature is depressed toward the ambient temperature, and water vapour and gases condense to form a liquid in the gathering header line. When the gas moves through a process facility, it is subjected to a number of changes, such as compression, temperature increase, cooling, and pressure reduction. As the processed gas moves through the transmission line to the point of use, there is a tendency for the gas within the pipe to attain the ambient temperature. Unless the gas has been properly dehydrated, condensation may occur.

The technology to be followed in handling the liquid contained in the gas starts at the point of collection (gathering) and continues through transmission in the gathering header system to the point of use, or process. The design engineer must install sufficient condensate collection points to drain the water and other liquids from the pipeline so that they do not interfere with gas flow.

A condensate collector (knockout pot or scrubber) should be provided for the gas immediately ahead of the compressor and also just before the use point. Condensate drains may be required in the gas transmission line, depending on its length, elevation, and the ambient temperatures encountered. If the gas moves through a simple blower system, there may be some additional condensation. To overcome this problem, the blower should be provided with a drain capability so that accumulated water will not interfere with fan operation. If, on the other hand, the gas enters a process station, a large knockout pot is usually provided. This collects condensate from gas cooled after discharge from the compressor, directing discharge back to the landfill or to other suitable points. This condensate will be acid in nature, and thus corrosive.

As liquids move through the process system, additional scrubber or knockout pots should be provided at points where liquids are expected to condense, and provisions should be made for their discharge in an environmentally safe manner.

Gas users will probably be concerned about the dryness of the gas as it reaches their facility. If the gas has not been depressed in temperature below the transmission temperature, then some water condensate and moisture surges may be expected at the point of use. In order to avoid the accumulation of moisture, the dew point of the gas must be lowered to a point below the ambient transmission line temperature and maximum transmission line pressure. To attain the required dew point, liquid can be removed at the processing station by utilizing either triethylene glycol and/or temperature depressing.

Methane Quality Improvement

Carbon dioxide is generated in the landfill in approximately the same percentage as methane (45 to 55%); therefore, one of the major efforts in upgrading methane gas quality is to separate the carbon dioxide from the methane.

A number of solvent treatment systems are available which utilize a liquid solvent that has an affinity for carbon dioxide, hydrogen sulphide and, in some instances, water. The solvent has minimal affinity for methane; thus the methane is effectively separated from the other gases. Solvent treatment systems currently in use include Methyl Ethanol Amine-Diethanol Amine Absorption (MEA-DEA), Diglycolamine (DGA), Hot Potassium Carbonate, Propylene Carbonate, Selexol, and Fluor Solvent.

Dry adsorbent systems can also be used where molecular sieve, activated charcoal or other appropriate adsorbents remove the contaminants. As an example, the molecular sieve has a microscopic honeycomb structure that traps (adsorbs) molecules according to their size and polarity. Some molecules, including carbon dioxide, hydrogen sulphide and water, are more readily adsorbed than others such as methane, thus allowing the landfill gas contaminants to be selectively removed. In all instances, the solvent or adsorbent is regenerated and recycled, the latter being regenerated through vacuum evacuation and/or thermal regeneration. The resulting contaminated gas or solvent is freed and discharged in an environmentally safe manner.

Each of the methane treatment processes should be evaluated on its own merit, with special consideration for the economics, environmental constraints and process reliability.

BIBLIOGRAPHY

1. Alpern, Robert, "Decomposition Rates of Garbage in Existing Los Angeles Landfills", unpublished Master's Thesis, California State University, Long Beach (1973).

2. Anderson, L.C., R.G. Kispert, S.E. Sadek, D.H. Walker, and D.L. Wise, "Fuel Gas Production from Solid Waste", Biotechnical and Bioengineering Symposium, No. 5, 285-301 (1975).

3. Augenstein, D.C., C.L. Cooney, R.L. Wentworth, and D.L. Wise, "Fuel Gas Recovery from Controlled Landfilling of Municipal Wastes", Resource Recovery and Conservation, 2, 103-117 (1976).

4. Blanchet, M.J., and staff of the Pacific Gas and Electric Company, "Treatment and Utilization of Landfill Gas", Mountain View Project Feasibility Study, EPA-530/SW-583, 1977.

5. Bowerman, F.R., N.K. Rohatgi, K.Y. Chen and R.A. Lockwood, A Case Study of the Los Angeles County Palos Verdes Landfill Gas Development Project, Ecological Research Series, EPA-600/3-047, July 1977.

6. Carlson, J.A., "Recovery of Landfill Gas at Mountain View", Engineering Site Study, EPA-530/SW-587d, May 1977.

7. Carruth, D. and A.J. Klee, "Sample Weights in Solid Waste Composition Studies", Journal of the Sanitary Engineering Division, ASCE, 96, 945 (1970).

8. Chan, D.B. and E.A. Pearson, Comprehensive Studies of Solid Wastes Management: Hydrolysis Rate of Cellulose in Anaerobic Fermentation, University of California, SERL Report No. 70-3, October 1970.

9. Committee on Sanitary Engineering Research, "Refuse Volume Reduction in a Sanitary Landfill", Journal of the Sanitary Engineering Division, ASCE, 85, 37 (1959).

10. Davis, S.N. and R.J.M. DeWiest, Hydrology, John Wiley, New York (1966).

11. DeWalle, F.B., E.S.K. Chian, and E. Hammerberg, "Gas Production from Solid Waste in Landfills", Journal of the Environmental Engineering Division, ASCE, 104, 415-432 (1978).

12. EMCON Associates and Jacobs Engineering Co., "A Feasibility Study of Recovery of Methane from Parcel I of the Scholl Canyon Sanitary Landfill", for the City of Glendale, California, October 1976.

13. Fair, G.M. and E.W. Moore, "Heat and Energy Relations in the Digestion of Sewage Solids", Sewage Works Journal, 430-439, May (1932).

14. Farquhar, G.H. and F.A. Rovers, "Gas Production During Refuse Decomposition", Water, Air and Soil Pollution, 2, 1973, 483-495.

15. Findikakis, A.N. and J.O. Leckie, "Numerical Simulation of Gas Flow in Sanitary Landfills", Journal of the Environmental Engineering Division, ASCE, submitted for publication, 1978.

16. Golueke, C.G. and P.H. McGauhey, Comprehensive Studies of Solid Waste Management, 1st and 2nd Annual Reports, Public Health Service Publication No. 2039 (1970).

17. Ham, R.K., W.K. Porter, and J.J. Reinhardt, "Refuse Milling for Landfill Disposal", Public Works, December 1971.

18. Hitte, S.J., "Anaerobic Digestion of Solid Waste and Sewage Sludge into Methane", reprinted from Compost Science, 17, No. 1 (January-February 1976).

19. Jones, B.B. and F. Owen, Some Notes on the Scientific Aspects of Controlled Tipping, City of Manchester, Report (1934).

20. Kaplovsky, A.J., "Volatile Acids Production During Digestion of Several Industrial Wastes", Sewage and Industrial Wastes, 24, 194 (1952).

21. Kaplovsky, A.J., "Volatile Acids Production During the Digestion of Seeded, Unseeded, and Limed Fresh Solids", Sewage and Industrial Wastes, 23, 713 (1951).

22. Katz, D.L., D. Cornell, R. Kobayashi, F.H. Poettman, J.A. Vary, J.R. Elenbaas and C.F. Weinaug, Handbook of Natural Gas Engineering, McGraw-Hill Book Co., N.Y. (1959).

23. Lawrence, A.W. and P.L. McCarty, "Kinetics of Methane Fermentation in an Anaerobic Treatment", Journal of the Water Pollution Control Federation, 41, R1-R17 (1969).

24. Leckie, J.O., "Estimation of Potential Methane Production from Solid Waste Landfills", unpublished report for EMCON Associates, September 1974.

25. Leckie, J.O., J.G. Pacey and C. Halvadakis, "Landfill Management with Moisture Control", Journal of the Environmental Engineering Division, ASCE, accepted for publication, 1978.

26. Los Angles, City of, Estimation of the Quantity and Quality of Landfill Gas from the Sheldon-Arleta Sanitary Landfill, Research and Planning Division, Bureau of Sanitation, City of Los Angeles, Technical Report (1976).

27. Louisville, Kentucky, Industrial Metropolitan Region Solid Waste Disposal Study, Vol. I., Jefferson County, Kentucky, Interim Report, USDHEW (1970).

28. McCarty, P.L., "Anaerobic Waste Treatment Fundamentals", Public Works, 95, No. 9-12 (1964).

29. McCarty, P.L., "Energetics and Kinetics of Anaerobic Treatment", Advanced Chemical Series, 105, 91-107 (1971).

30. McCarty, P.L., J.S. Jeris and W. Murdoch, "Individual Volatile Acids in Anaerobic Treatment", Journal of Water Pollution Control Federation, 35, 1501-1516 (1963).

31. McCarty, P.L., "Stoichiometry of Biological Reactions", Proceedings of the International Conference Towards a Unified Concept of Biological Waste Treatment Design, Atlanta, Georgia (1972).

32. McCarty, P.L., "Stoichiometry of Biological Reactions", Progress in Water Technology, 7, 157-172 (1975).

33. McCarty, P.L., "Thermodynamics of Biological Synthesis and Growth", Proceedings of the Second International Water Pollution Research Conference (Tokyo), 169-187, Pergamon Press (1965).

34. Ministry of Housing and Local Government, Pollution of Water by Tipper Refuse, Report of the Technical Committee on the Experimental Disposal of House Refuse in Wet and Dry Pits, H. M. Stationery Office (1961).

35. Mohsen, F.N.M., G.J. Farquhar and N. Kouwen, "Modelling Methane Migration in Soil", Journal of the Environmental Engineering Division, ASCE, submitted for publication, 1977.

36. Mohsen, F.N.M., "Gas Migration from Sanitary Landfills and Associated Problems", Ph.D. Thesis, University of Waterloo, Waterloo, Ontario, Canada, 1975.

37. Monod, J., "La Technique de Culture Continue: Theorie et Applications", Annals Institute Pasteur, 79, 390-410 (1950).

38. Moore, C.A., "Theoretical Approach to Gas Movement Through Soils", Progress Report on EPA Contract No. 68-03-0326.

39. National Center for Resource Recovery, "Municipal Solid Waste: Its Volume, Composition and Value", NCCR Bulletin, III, (2) 4- (1973).

40. National Center for Urban and Industrial Health, Elements of Solid Waste Management, Training Course Manual in Solid Wastes, NCUIH, Cincinnati (1968).

41. New York University, Landfill Survey, Sanitary Engineering Research Laboratory Report (1940).

42. Pfeffer, J.T., Reclamation of Energy from Organic Waste, Environmental Protection Technology Series, EPA-670/2-74-016, March 1974.

43. Quad-City Solid Wastes Project, Interim Report, June 1, 1966 to May 31, 1967, Cincinnati, USDHEW (1968).

44. Reid, R.C. and T.K. Sherwood, The Properties of Gases and Liquids, McGraw-Hill, New York (1966).

45. Schwegler, Ronald, "Energy Recovery at the Landfill", Seminar of the Governmental Refuse Collection and Disposal Association, Santa Cruz, California, November 7-9, 1973.

46. Shuster, W.W., Partial Oxidation of Solid Wastes, U.S. Public Health Service, Publication No. 2133 (1970).

47. Solid Waste Management, "Composition of Rubbish in the United States", Solid Waste Management, 15, (9) 75 (1972).

48. Sonoma County Department of Public Works and EMCON Associates, Sonoma County Refuse Stabilization Study, Third Annual Report, Project G06-EC-00351, July 1974.

49. State of California, Effects of Refuse Dumps on Groundwater Quality, California State Water Pollution Control Board, Publication No. 24 (1961).

50. State of California, In-Situ Investigation of Movements of Gases Produced from Decomposing Refuse, California State Water Quality Control Board, Publication No. 31 (1965).

51. Tchobanoglous, G., H. Theisen, and R. Eliassen, Solid Wastes: Engineering Principles and Management Issues, McGraw-Hill Book Co., N.Y. (1977).

52. Todd, D.K., Groundwater Hydrology, John Wiley and Sons, Inc., 1959.

53. University of California, Reclamation of Municipal Refuse by Composting, Sanitary Engineering Research Project, Technical Bulletin No. 9 (1953).

54. U.S. EPA, Fourth Report to Congress: Resource Recovery and Waste Reduction, U.S. EPA Report SW-600 (1977).

55. Walton, W.C., Groundwater Resource Evaluation, McGraw-Hill Book Company (1970).

56. Wise, D.L., S.E. Sadek, R.G. Kispert, L.O. Anderson and D.H. Walker, "Fuel Gas Production from Solid Waste", Biotechnology and Bioengineering Symposium, No. 5, 285-301, John Wiley & Sons, Inc. (1975).

APPENDICES

APPENDIX A

LIST OF ABBREVIATIONS

BOD	biochemical oxygen demand
Btu	British thermal unit
^{o}C	degree Celsius
cfm	cubic foot per minute
cm	centimetre
C/N ratio	weight of carbon ÷ weight of nitrogen in sample
COD	chemical oxygen demand
DGA	diglycolamine
^{o}F	degree Fahrenheit
ft	foot
ft^2	square foot
ft^3	cubic foot
g	gram
in	inch
kg	kilogram
kJ	kilojoule
kN	kilonewton
L	litre
lb	pound (mass)
lb_f	pound (force)
LEL	lower explosive range
m	metre
m^2	square metre
m^3	cubic metre
MEA-DEA	methyl ethanol amine-diethanol amine absorption
mg	milligram
mL	millilitre
mm	millimetre

N	newton
PG&E	Pacific Gas and Electric Company
Psig	pounds per square inch
PVC	polyvinyl chloride
scf	standard cubic foot
typ.	typical
yd	yard

APPENDIX B

LIST OF VARIABLES AND SYMBOLS[a]

Pages 5-13

pH	-log $[H^+]$, where $[H^+]$ is the concentration of hydrogen ions, H^+
CH_4	methane
CO_2	carbon dioxide
$C_5H_7O_2N$	chemical formula for a bacterial cell
s	fraction of waste chemical oxygen demand (COD) synthesized or converted to cells
e	fraction of waste COD converted to methane gas for energy
θc	solids retention time, days
a_e	S_{max} when $\theta c = 0$
f	cell decay rate, day^{-1} (per day)
Y	biological yield factor, generally ranging from 0.03-0.15
HAc	acetic acid
HCO_3^-	bicarbonate ion
BA	bicarbonate alkalinity, mg/L as $CaCO_3$
TA	total alkalinity, mg/L as $CaCO_3$
VA	volatile acids, mg/L as acetic acid (HAc)
Σx	(garden waste + paper products + textiles + wood) on a dry weight basis
$C_{203}H_{334}O_{138}N$	empirical formula for paper
$C_{16}H_{27}O_8N$	empirical formula for food wastes

[a]In the technical literature one variable (letter) is occasionally used to represent more than one parameter. For example, in one chapter r refers to the radial distance from recovery well, while in another r is the lower limit on sustainable methane recovery rate for the entire site. With the goal of minimizing confusion, this list is organized by chapter (page numbers given), in the order in which the variables occur in the text. Where appropriate, the reference equation is included.

Pages 25-36

C_i	$k \times k \times W_t \times P_i \times (1 - M_i) \times V_i \times E_i$
W_t	weight total bulk refuse
P_i	fraction of component i by wet weight of total bulk wet refuse
M_i	fractional moisture content of component i by wet weight
V_i	fractional volatile solids content of total bulk dry weight
E_i	fraction of biodegradable volatile solids of component i
k	351 L CH_4/kg S_i
k'	1.5 kg COD/kg
C_t	$\Sigma_i C_i$

Pages 37-58

S	concentration of substrate surrounding the microorganisms
K	maximum rate of substrate utilization per unit mass of microorganisms
X	concentration of microorganisms
K_s	half-velocity coefficient
t	time
G	volume of gas produced prior to time t
G_0	G at t=0
L	volume of gas remaining to be produced after t
L_0	L at t=0
k_1	first stage gas production rate constant
k_2	second stage gas production rate constant
$t_{1/2}$	half-time, when half of the ultimate gas production has been reached, $G = L = L_{0/2}$ when $t = t_{1/2}$
$t_{99/100}$	time required to achieve 99% of ultimate gas production
t_{max}	time of maximum production rate
k	gas production constant for the Scholl Canyon Kinetic Model
n	number of submasses considered
k_i	gas production constant for submass i

v_i	fraction of total refuse mass contained in submass i
t_i	time from placement of submass i to point in time at which composite production rate is desired

Pages 59-64

R	Reynolds number
μ	absolute viscosity of the fluid
ρ	density of the fluid
υ	velocity of flow
D	characteristic dimension of the system or critical grain diameter
r	radial distance from the recovery well
v_r	apparent gas velocity at r in a direction toward the well
k	coefficient of permeability
h	static head (piezometric head)
p	total pressure at distance r
γ	specific weight of the gas
z	elevation above some arbitary datum
K	standard coefficient of permeability
k_s	specific or intrinsic permeability of the medium

Pages 65-95

Q_w	well-flow rate
K	compilation of conversion factors
R	radius of influence
t	refuse thickness
D	in-place refuse density
r	methane production rate (where $Q_w = \frac{K R^2 + Dr}{C}$)
C	fractional methane concentration
I	influence or relative pressure drop
r	radial distance from the well being pumped (where $I = A + B\ln r$)
A	a positive constant
B	a negative constant
Q	magnitude of the gas flow rate across imaginary cylindrical surface

Q^*	optimal flow rate
R^*	radius of influence at the rate Q^*
r	radius of the imaginary cylindrical surface (where $Q = \pi(R^2 - r^2)tDFg$)
h	depth of well
F_g	gas production rate per unit mass of refuse
Q	lower limit on sustainable methane recovery rate for the entire site (where $Q = \frac{(X-V)C}{fp}$)
r	lower limit on sustainable methane recovery rate per unit mass of refuse (where $r = \frac{(X-V)C}{fM(P/365)}$)
x	total volume of gas extracted
V	landfill void volume
C	methane concentration (% ÷ 100)
P	duration of extraction testing
M	refuse mass
f	fraction of in-place refuse from which the gas was extracted
Q	volumetric flow rate (where $\upsilon_r = \frac{Q}{A}$)
r_w	radius of the well
p_w	pressure at the well/refuse interface
p_e	pressure at the "radius of influence" of the well

APPENDIX C

CONVERSION FACTORS

1 centimetre	0.3937 inches
1 centimetre	3.281×10^{-2} feet
1 cubic metre	1000 L
1 degree Celsius	0.556 ($^{\circ}$f - 32) degrees
1 dyne per square centimetre	9.869×10^{-7} atmospheres
1 dyne-second per square centimetre	1.00 poise
1 hectare	2.471 acres
1 kilogram	2.205 pounds (mass)
1 kilogram	1.102×10^{-3} tons
1 kilogram per cubic metre	6.243×10^{-2} pound (mass) per cubic foot
1 kilogram per cubic metre	1.686 pound (mass) per cubic yard
1 kilojoule	0.9481 British thermal units (Btu)
1 kilojoule per cubic metre	0.02684 Btu per cubic foot
1 kilojoule per kilogram	0.4301 Btu per pound (mass)
1 kilojoule per litre	26.84 Btu per cubic foot
1 kilonewton per square metre	0.1450 pounds per square inch
1 litre	3.532×10^{-2} cubic feet
1 litre per kilogram	0.0160 cubic foot per pound (mass)
1 litre per second	2.119 cubic feet per minute
1 metre	3.281 feet
1 metre	1.0936 yards
1 millimetre of mercury	0.5353 inches of water (at 4°C)
1 millimetre per kilogram per day	5.85×10^{-3} cubic feet per pound per year
1 newton	0.2248 pounds (force)
1 newton-second per m^2	2.089×10^{-2} pound-sec per square foot

1 poise	100 centipoise
1 square centimetre	1.013×10^{8} darcy
1 square centimetre	1.076×10^{-3} square feet
1 square foot	9.416×10^{10} darcy
1 square metre	10.76 square feet

APPENDIX D

SYNOPSIS OF RECENT RESEARCH IN ANAEROBIC DIGESTION*

Research on anaerobic processes involving the biodegradation of decomposable complex organics in the last decade has revealed a complex biochemical process usually termed anaerobic digestion (Zehnder 1978). The end products produced by anaerobic digestion consist primarily of methane and carbon dioxide and water. Also produced are trace amounts of other substances, including hydrogen and hydrogen sulphide gases, as well as dissolved ammonium and bicarbonate ions.

Three physiologically distinct groups of microorganisms are involved in converting complex organics to methane. The activity of these three microbial groups gives rise to the three steps in the overall process of anaerobic digestion (shown on Figure D-1).

In the first step, complex organic materials, carbohydrates, proteins and lipids are hydrolyzed and fermented to fatty acids, alcohols, carbon dioxide, ammonia, and some hydrogen. The organisms involved in this step are anaerobic and facultative anaerobic fermenting bacteria such as *Bacillus*, *Clostridium*, and enterobacteria.

In the second step of anaerobic digestion, organic acids and alcohols produced during the first step are converted into acetate, hydrogen, and carbon dioxide. This process is controlled by the actual hydrogen partial pressure, since organisms in this step have to catalyze reactions which are endothermic under standard conditions at a pH of 7. Very little is known about these organisms which exist only in co-cultures with other organisms which keep the hydrogen partial pressure low, e.g., methane bacteria.

The third step involves two physiologically distinct types of methanogenic bacteria: (1) methanogens that reduce carbon dioxide to methane; and (2) methanogens that decarboxylate acetate to methane and carbon dioxide.

---

* Based on the work of A.J.B. Zehnder in "Ecology of Methane Formation", Water Pollution Microbiology, 2, 349-376 (1978).

In the case of sewage sludge digestion, about 70% of the methane produced has been found to originate from acetate, while the remaining 30% is derived from carbon dioxide reduction (Zehnder 1978).

The following equations summarize the three steps for the conversion of glucose ($C_6H_{12}O_6$) to methane and carbon dioxide (Zehnder 1980)*.

Step 1

Fermentation $C_6H_{12}O_6 \quad 2CO_2 + 2C_2H_5OH$

Step 2

Hydrogen Formation $2C_2H_5OH + 2H_2O \quad 2CH_3COOH + 4H_2$

Step 3

CO_2 Reduction $CO_2 + 4H_2 \quad CH_4 + 2H_2O$

Acetate Decarboxylation $2CH_3COOH \quad 2CH_4 + 2CO_2$

Overall Reaction $C_6H_{12}O_6 \quad 3CH_4 + 3CO_2$

---

* Zehnder, A.J.B., "Water Microbiology", Class Notes, Stanford University (1980).

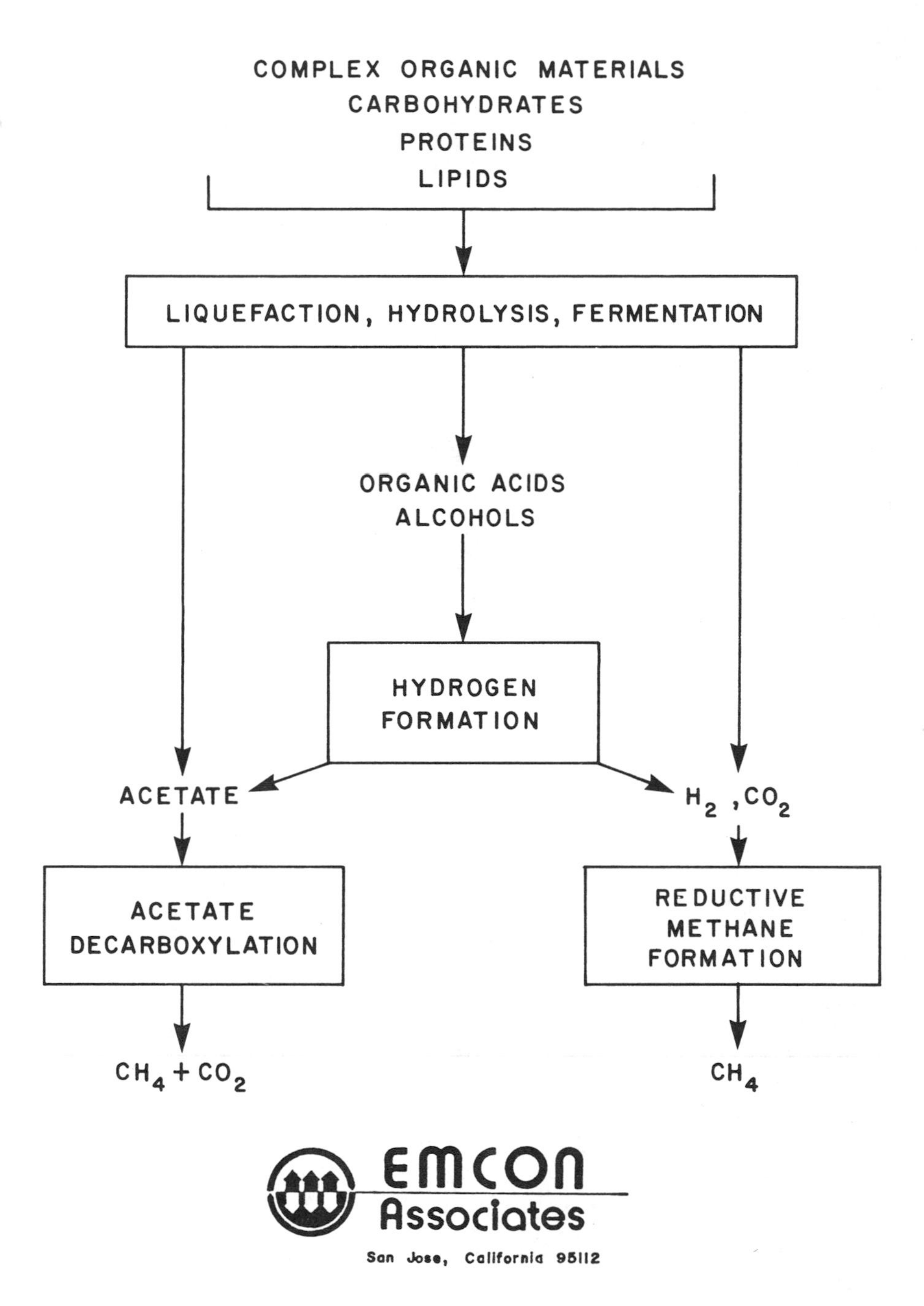

Figure D-1. Multistep methanogenesis in anaerobic digestion. (Source: Zehnder 1978.)

INDEX